FRANCIS SANDBACH

Principles of pollution control

Longman
London and New York

Longman Group Limited
Longman House
Burnt Mill, Harlow, Essex, UK

*Published in the United States of America
by Longman Inc., New York*

© Longman Group Limited 1982

All rights reserved. No part of this publication may be reproduced, stored in a retrieval system, or transmitted in any form or by any means, electronic, mechanical, photocopying, recording, or otherwise, without the prior permission of the Copyright owner.

First published 1982

British Library Cataloguing in Publication Data

Sandbach, Francis
  Principles of pollution control. – (Themes in resource management)
  1. Pollution
  I. Title   II. Series
  628.5   TD178

  ISBN 0-582-30042-8

Library of Congress Cataloging in Publication Data

Sandbach, Francis.
  Principles of pollution control.

  (Themes in resource management)
  Bibliography: p.
  Includes index.
  1. Pollution. 2. Factory and trade waste.
I. Title.   II. Series.
TD194.S26   363.7'3   81-11718

ISBN 0-582-30042-8 (pbk.)   AACR2

Printed in Great Britain by William Clowes (Beccles) Ltd
Beccles and London

ITEM NO: 2025420

**UNIVERSITY OF GLAMORGAN
LEARNING RESOURCES CENTRE**

Pontypridd, Mid Glamorgan, CF37 1DL
Telephone: Pontypridd (0443) 480480

Books are to be returned on or before the last date below

ONE WEEK LOAN

25. OCT 1994	15. JAN 96
18. MAR 95	03. MAY 1996
	10 MAR 1997
13. OCT 95	
15 DEC 1995	
18. OCT 9.	
27. OCT	

THEMES IN RESOURCE MANAGEMENT
Edited by Professor Bruce Mitchell, University of Waterloo

Francis Sandbach: Principles of pollution control
R. L. Heathcote: Arid Lands: Their Use and Abuse
Stephen Smith: Recreation Geography
John Blunden: Mineral Resources and their Management
Paul Eagles: The Planning and Management of Environmentally Sensitive Areas
John Chapman: A Geography of Energy

Contents

List of Figures	viii
List of Tables	ix
Acknowledgements	x
Foreword	xi
Introduction	xiii

Chapter 1	The pollution problem	1
	Introduction	1
	The pollution concept	2
	Different pollution control strategies	5
	Economic efficiency	10
	Estimation of pollution damage – the dose/response relationship	11
	Different types of standard	14
	Standards based on threshold limits	16
	Revealed preferences	17
	Expressed preferences	18
	Cost-benefit analysis	19
	Environmental impact assessment	20
	Conclusions	21
Chapter 2	Pollution hazards: the problems of uncertainty and risk	23
	Hazard management	23
	Uncertainties of risk	25
	The surprising, but conceivable accident	26
	Nuclear power	28
	Balancing risks	31
	The politics of uncertainty and the monopoly of information	34
	Conclusions	35

v

Contents

Chapter 3	Prevention rather than cure: an alternative strategy	36
	Recycling	37
	Economic and organizational constraints on recycling	39
	Product substitution, product use and the social organization of production – the case of China	40
	Alternative technology	42
	Conclusions	47
Chapter 4	International problems of pollution control	48
	Transfrontier pollution	48
	International agreements	51
	Mobile sources and mobile products	53
	Transportation of oil	53
	Mobile products: the case of PCBs	56
	Regional and global pollution problems	56
	Trade and economic relationships	57
	Conclusions	59
Chapter 5	Institutional and administrative arrangements for pollution control	61
	Introduction	61
	Administrative arrangements in the UK	62
	International comparison of pollution control policy-making and administration	63
	Centralization versus decentralization	68
	The EEC's approach to pollution control	73
	Conclusions	76
Chapter 6	Economic considerations in pollution control and the regulatory approach in practice	77
	Introduction	77
	The advantages of pollution taxes	80
	The disadvantages of pollution taxes	85
	Features of the regulatory approach	90
	Enforcement problems	90
	The American permit system	92
	The British consent system	93
	Industrial air pollution control in Britain	94
	Conclusions	101
Chapter 7	Public participation and economic interests in pollution control	102
	The pluralist perspective	102
	The historical roots of the British Clean Air Act 1956	104

	The Alkali Act 1863	115
	Economic interests in pollution control decision-making in the post-war period	118
	Conclusions	122
Chapter 8	Pollution in the asbestos industry	124
	Introduction	124
	History of manufacture and use	125
	Asbestos dust conditions from mining to disposal of waste	128
	The hazards of asbestos revealed – an historical perspective	130
	Influence of industrial capital on knowledge and understanding of risks	135
	Asbestos regulations in the UK and elsewhere and the dose/response relationship	139
	Problems of enforcement	148
	Conclusions	153
Chapter 9	Conclusions	155
	Bibliography	160
	Index	172

List of figures

1.1	Chronic bronchitis: distribution of mortality in the UK, males 1959–63	4
1.2	Sulphur dioxide: trends in emissions and average urban concentrations	8
1.3	Typical asymptotic thresholds or time/concentration curves	12
2.1	Hazard causation and possible stages of intervention	24
3.1	The self-sufficient home	43
3.2	The alternative mistopia	46
5.1	Responsibilities for pollution control in England and Wales	64
5.2	Processes involved in the making of a directive	70
6.1	The optimum level of pollution control without threshold	78
6.2	The optimum level of pollution control with threshold	80
6.3	The innovative incentive of a pollution tax	81
6.4	The allocative advantage of a pollution tax	83
7.1	Fogs at Greenwich	107

List of tables

1.1	Path of sulphur oxides from oil	9
2.1	Rasmussen's estimate of possible consequences of 'An Extremely Serious Accident'	28
2.2	Dose rates in the UK from ionizing radiation	33
3.1	Western-world recycling ratios	38
4.1	Sources of oil in the marine environment	54
6.1	Alkali Act prosecutions	97
7.1	Consumption of coal in the UK according to use	109
8.1	Trends in world production of asbestos	124
8.2	Estimated world production of asbestos by country and region, 1979	125
8.3	Imports of asbestos to the UK	126
8.4	Estimated annual sales turnover for asbestos industry	127
8.5	Asbestos distribution by end use, grade and type in the US, 1974	128
8.6	Concentrations of asbestos dust measured during operations on asbestos cement	129
8.7	Expected and observed deaths among 632 New York and New Jersey asbestos-workers exposed to asbestos dust twenty years or longer (1 January 1943 to 31 December 1971)	134
8.8	Present hygiene standards for asbestos in fifteen countries	136
8.9	Death certificates mentioning mesothelioma, etc. UK, 1967–76.	138
8.10	Excess of observed minus expected deaths. Based on 8,327 deaths from industrial cohorts reported in eleven studies	139
8.11	Fibre reinforcement of cement	146
8.12	Impact on asbestos industries of 1 fibre/ml limit: increases in costs, prices, exports, imports, sales and employment	149

Acknowledgements

I should like to thank Philip Lowe of University College, London, and Bruce Mitchell, the series editor and professor at the University of Waterloo, for constructive comments upon the draft manuscript. I owe a debt to past students who have made it possible for me to pursue an interest in pollution problems. I am extremely grateful to my father who has helped to make the book more readable. Alison Bowd has translated numerous near illegible drafts into readable typescript prior to publication, a task which most authors rightly recognize as invaluable. Finally I should like to thank my wife and family for tolerating the antisocial behaviour so often unavoidable in writing.

Francis Sandbach
March 1981

We are grateful to the following for permission to reproduce copyright material:

Asbestos Information Centre for Our Tables 8.4, 8.11 and 8.12; Central Directorate on Environmental Pollution for Our Fig. 5.2 (Renshaw 1978) by permission of the Controller of Her Majesty's Stationery Office; Graham and Trotman Ltd for our Table 4.1 (Wardley-Smith 1979 II); Heldref Publications for our Fig. 2.1 (Fischoff et al 1978a) and Our Table 8.7 (adapted from Ahmed et al 1972); Her Majesty's Stationery Office for Our Fig. 1.2 (Dept of Environment 1980), Our Table 1.1 p 3 (Marquand 1979) *Digest of UK Energy Statistics* 1975 and *National Society for Clean Air Yearbook* 1975, Our Table 2.2 (Royal Commission on Environmental Pollution 1976b), Our Table 8.3 (adapted from Advisory Committee on Asbestos 1979b), Our Table 8.6 and 8.10 (Advisory Committee on Asbestos 1979a), Our Table 8.8 (Advisory Committee on Asbestos 1979b), Our Table 8.9 (Health and Safe-

Acknowledgements

ty Executive 1977), Our Fig. 5.1 (Central Directorate on Environmental Pollution) by permission of the Controller of Her Majesty's Stationery Office; Heyden and Son Ltd for an extract from 'The Historical Roots of the British Clean Air Act 1956' by Francis Sandbach *Interdisciplinary Science Reviews* Vol 6 No 3 (c) Heyden and Son Ltd 1981, Our Fig. 7.1 (Ashby and Anderson 1977 III); the author, Professor G. Melvyn Howe for Our Fig. 1.1 (Howe 1972); International Agency for Research on Cancer for our Table 8.1 from Table 8 IARC Monographs on the Evaluation of the Carcinogenic Risk of Chemicals to Humans Vol 14 International Agency for Research On Cancer 1977; Longman Group Ltd for Our Table 3.1 (Pearce 1976c); Manchester University Press for Our Fig. 1.3 (Saunders 1976); Observer News Service for Our Table 2.1 (Bugler 1979); United States Department of the Interior for Our Table 8.5 (Clifton 1974).

Foreword

The 'Themes in Resource Management' series has several objectives. One is to identify and to examine substantive and enduring resource management and development problems. To meet this end, books have been commissioned on a variety of problem areas deemed to be significant in the years ahead. In this first book in the series, Francis Sandbach has focused upon pollution problems, environmental health and the issues of risk and uncertainty in resource management. Other books will address problems related to energy, minerals, water, arid lands, environmentally sensitive areas and recreation. Attention will range from local to international scales, from developed to developing nations, from the public to the private sector, and from biophysical to political considerations.

A second objective is to assess responses to these management and development problems in a variety of world regions. Several responses are of particular interest: *research* and *action programmes*. The former involves the different types of analysis which have been generated by natural resource problems. The series will assess the kinds of problems being defined by investigators, the nature and adequacy of evidence being assembled, the kinds of interpretations and arguments being presented, the contributions to improving theoretical understanding as well as resolving pressing problems, and the areas in which progress and frustration are being experienced. The latter response involves the policies, programmes and projects being conceived and implemented to tackle complex and difficult problems. The series is concerned with reviewing their adequacy and effectiveness in terms of economic efficiency, social equity and environmental harmony. This orientation should provide a departure point for considering the transferability of experience from one jurisdiction to another.

Research and action programmes were discussed above as separate responses. However, a third objective is to explore the way in which resource analysis, management and development might be made more complementary to one another. Too often analysts and managers go their separate ways. A good part of the blame for this situation must lie with the analysts who too frequently ignore or neglect the concerns of managers, unduly emphasize method and technique, and exclude explicit consideration of the managerial implications of their research. It is hoped that this series will demonstrate that research and analysis can contribute both to theoretical development and to resolution of important problems.

Francis Sandbach's book on principles of pollution control is most appropriate as the first book in this series. The book has a solid conceptual foundation combined with empirical testing of concepts and techniques in pollution control. The theories, concepts and findings of a variety of disciplines are integrated and appraised. While the majority of the examples and the case study of asbestos pollution are drawn from Britain, numerous other examples from Europe, North America and Asia facilitate international comparisons.

Sandbach explicitly identifies the constraints under which policy-makers operate, and explores the trade-offs and compromises which are, could be, and should be made. While sensitive to the environment within which managers function, he also focuses incisively upon research accomplishments and short-comings. Hence, he notes the inadequacy in coming to grips with such basic tasks as defining and measuring pollution He also suggests that researchers are not always altruistic in their work, and have too often allowed their choice of research design and their interpretations of evidence to be influenced by vested interests. This point raises major ethical questions about the way in which research is conducted.

Sandbach explores alternative responses, and in this manner is a constructive critic rather than simply a critic of both researchers and managers in the complex field of pollution control. Throughout, he stresses the overriding importance of risk and uncertainty in resource management, a facet which is often discussed in general terms rather than in the context of specific management problems. The concern with risk and uncertainty, and the attention to environmental health as revealed by the examination of the asbestos industry, suggests that this book is addressing the resource management needs of the 1980s and 1990s.

Bruce Mitchell
Visiting Professor
University of Leeds

March 1981

Introduction

Knowledge of pollution has increased in leaps and bounds. Our understanding of pollution has developed a long way from the days when smoke was a sign of prosperity, and the pea-souper was the romantic setting for Sherlock Holme's mysteries. What are considered pollutants today were not necessarily considered such in the past. Fifty years ago arsenic was the only metal known to be a carcinogen. Today numerous metals and substances are known to be causes of cancer. However, it would be complacent to believe that we are now dealing satisfactorily with all pollutants.

There are about 30,000 chemicals made in quantities greater than one metric ton per annum. Little is known about the possible danger to human health and environmental resources from the majority of these chemicals. Yet what little is known cannot give us any grounds to believe that we have contained the pollution problem. For instance, it is known that of all the major causes of death in industrialized countries only cancer has continued to rise year after year throughout this century. In the US it has been estimated that, after adjustment for age, the incidence (new cases) of cancer has been rising at 1.6 per cent per annum and that as many as 80 to 90 per cent of cancers are caused by environmental factors. About 10 per cent of 7,000 substances that have been specifically tested for carcinogenicity have been shown to be carcinogens (Council on Environmental Quality 1979: 188–206).

Before any action against a pollutant can take place it must be identified, its extent estimated, the problem evaluated, and then finally the pollutant controlled. This whole process may be regarded superficially as solely a technical matter, but whether, how and when this process occurs is a matter of political and economic interest. It is the purpose of this book to reveal the main influences on these different aspects of pollution control policy as well as to discuss the merits of a wide variety of strategies and institutional arrangements. The latter must, however, be assessed in con-

Introduction

junction with the former. Institutions and techniques of pollution control operate within the confines of broader social influences, influences that reflect and reinforce the organization of the economy and society.

Inevitably the scope of this book will involve a confrontation of different theories and perspectives on pollution control. For example, the economic debate over the merits of pollution taxes versus a standards/enforcement strategy leads one to question whether or not polluters, administrators and pressure groups operate in the way that is assumed by the economist. Any rules laid down for achieving economically efficient solutions to pollution problems must be seen in the light of economic and political interests bent on redefining the rules in their own interest. The theoretical arguments of economists, and there is now a large literature on this area, have suffered in the past from a lack of empirical support and so have often been too abstract and vacuous. In this book emphasis is placed upon case material to support or refute theoretical insights. Lack of critical empirical insights has unfortunately allowed a considerable amount of dubious theorizing to flourish. Part of the task of this book is to point to the limitations of these types of analysis.

The sequence of chapters in this book is designed to lead the reader from fairly straightforward descriptions and categorizations of problems to a greater understanding of the economic and political influences on pollution control policy. The former is considered necessary groundwork on which to base a firmer understanding of the latter. So in the first chapter we start with problems concerning the definition of pollution, different control strategies, estimation of damage, and risk/benefit/cost analysis. Chapter 2 looks at the problems of uncertainty and risk of pollution in economic development and the scope for the mobilization of bias where uncertainty exists. In Chapter 3 the possibilities of preventing pollution through recycling and alternative technology are examined in the belief that prevention is better than cure, an axiom that is all too absent in today's pollution policy literature. Some of the constraints on alternative non-polluting strategies of economic development are examined.

Chapter 4 looks at the international characteristics of the pollution problem. Again at this stage much of the discussion is concerned with defining the nature of the problem and the characteristics of the international response to date. Similarly, Chapter 5 reviews institutional arrangements within nations and the EEC. Different types of administration give rise to different types of pollution problems. In Chapter 6 we can move on to look at the theoretical debate over whether pollution taxes or pollution standards lead to greater efficiency. Case studies of their administrative operation suggest that it is necessary to look more closely at the wider influence of pressure groups and economic interests. Chapter 7 assesses the importance of public participation and interest groups in dealing with pollution. Case studies are again used to illustrate the limitations and strengths

Introduction

of alternative perspectives. Finally Chapter 8 builds upon all the previous chapters by looking at one detailed example of pollution in the asbestos industry. This chapter is offered as an example of how one should go about investigating the characteristics of a pollution problem and the influences affecting it.

CHAPTER 1

The pollution problem

Introduction

During the last two decades we have witnessed considerable public interest in pollution and it has become a major political issue, especially in developed industrial nations. The *Torrey Canyon* oil spill, Minamata mercury poisoning, pesticide damage to wildlife and other dramatic incidents connected with pollution were widely publicized. Governments of the day reacted by setting up committees of inquiry, by increasing expenditure on pollution control, and by introducing legislation. For example, in the US the Clean Air Act of 1970 and the Clean Water Act of 1972 were major commitments to reducing pollution. Despite such measures the pollution problem remains with us today. To picture the full extent of this problem would necessitate resorting to voluminous information, but a few examples suffice to show that it is still a most serious one.

In the case of some pollutants, control measures and economic development have resulted in a reduction in pollution. Levels of smoke emission in the UK have fallen by 80 per cent between 1960 and 1977. Lead and arsenic emissions have also decreased between 1973 and 1976, but the same is not true of carbon monoxide, hydrocarbons and nitrogen oxides, common pollutants from motor cars. Over the last twenty years the level of noise, particularly in urban areas, has risen substantially, mainly because of traffic.

Some regional and global problems have also become more serious. Acid rains containing oxides of sulphur and nitrogen have increased in pace with the use of fossil fuels, especially in power plants, motor vehicles, and in the smelting of non-ferrous metals. Sulphur emissions in Europe increased by 150 per cent between 1950 and 1975. In the highly industrialized areas of the north-eastern US levels of acid in rainfall are now twenty times higher than they were in 1955 (Rosencranz and Wetstone 1980). To some extent the acid rain problem is local in character. The air of a heavy industrial city such as Chicago (US) has a mean annual concentration of sulphur dioxide some 1,300 times higher than an unpolluted area like Hawaii (Wainwright 1980). Nevertheless, despite local variations oxides of sulphur and nitrogen are highly mobile and cause regional and international problems.

Published estimates in 1979 indicated that the US exported about four million metric tons of sulphur dioxide pollution to Canada each year. Canada in its turn exports about a quarter of this amount to the US (Wetstone 1980). Data from 1973 for Western Europe indicated that an average 63 per cent of sulphur emissions represented national depositions, 17 per cent reciprocal depositions, and 20 per cent one-way depositions. The Scandinavian countries suffered in particular from transfrontier pollution. In Sweden and Norway at least 56 per cent of sulphur deposits come from other countries (Rosencranz 1980). Sweden has placed the economic loss of its recreational and commercial fishing industry due to acid rain at between $50 m. and $100 m. in 1973 (Council on Environmental Quality 1979: 71). This problem will be dealt with more fully in Chapter 4.

For many less conventional pollutants the nature, extent and damage caused by them is less certain. In this chapter we will look at the concept of pollution itself, an important prerequisite to further study. The concept of pollution allows us to define different pollution control strategies which leads us on to consider what types of information are needed for developing control measures. Finally we can look at how pollution damage is estimated and how pollution costs and risks can be balanced against benefits of control.

The pollution concept

It is worthwhile looking at the meaning of our subject in a little detail so as to avoid attributing disagreeable changes in the environment to something other than pollution. The common definition of pollution usually requires that the pollutant has caused a change in the environment and that this change is harmful. The notion of what is harmful has changed over time to incorporate interests other than those of immediate human concern. The following definition from an OECD (1976a) document on 'principles concerning transfrontier pollution' is typical of the general wording which has formed the starting point for international debate on pollution control:

> Pollution means the introduction by man, directly or indirectly, of substances or energy into the environment resulting in deleterious effects of such a nature as to endanger human health, harm living resources and ecosystems, and impair or interfere with amenities and other legitimate uses of the environnment.

It is necessary to stress human activity in the definition of pollution, for disagreeable changes in the environment may also result from natural hazards or 'natural' changes in the environment. Disappearance of fish, for example, may be due to overfishing or pollution, but it could also be due to changes in sea temperature and climate. It is obviously important here to distinguish between 'natural' causes and 'human' causes. The biologist, when investigating whether a river is polluted or not, must examine the biota (plants and animals) of that river in the light of what could reasonably be expected if there were no pollution. Changes in the biota of the river will depend not just upon the discharge of pollutants, but upon changes in other conditions such as the nature of stream bed and banks, temperature and turbidity of the water, and so on. If one travels down a river in its natural state one would expect to see changes in the biota of the river when physical changes in the river occur. The introduction of a pollutant will change the conditions and affect the biota that would normally be expected.

It often is difficult to assess the effect of all the interacting factors. Over recent decades, there have been changes in minute organisms in the Atlantic, but whether this has been due to climatic changes, pollution or some other factor is not known (Central Unit on Environmental Pollution 1974: 6). The importance of distinguishing between 'natural' causes and human pollution is also seen in the study of environmental health. Howe (1972) has shown how the incidence of disease varies as a consequence of environmental factors. The distribution of chronic bronchitis, for example, suggests that it is a town disease (Fig. 1.1). It appears much more frequently in smoke-polluted atmospheres. Nevertheless, although there is a clear correlation between bronchitis and smoke-induced fogs, which occur more often in polluted areas, bronchitis and respiratory diseases may also be affected by natural changes in the climate such as changes from warm to cold air or the natural formation of fogs. Consequently, variations in the incidence of bronchitis are also influenced by geographical differences in the 'natural' climate. In the case of some pollutants there is a very close correlation between distribution of pollution and environmental characteristics. Lichens, for example, are extremely sensitive to air pollution, especially sulphur dioxide, and their distribution in England and Wales follows very closely the distribution of air pollution (Hawkesworth and Rose 1970).

When talking about pollution it is important to consider 'actual harm'. The impact of pollution is not uniform, and clearly has different effects on different targets (targets refer to human recipients, plants, animals and physical elements of the ecosystem). Health consequences, for example, may depend upon people's ages. The notorious London smog of 1952 led to about 4,000 more deaths than average, but these deaths occurred mainly in adults already suffering from respiratory and heart diseases. High nitrate concentrations in water are most dangerous for infants, for they cause

The pollution problem

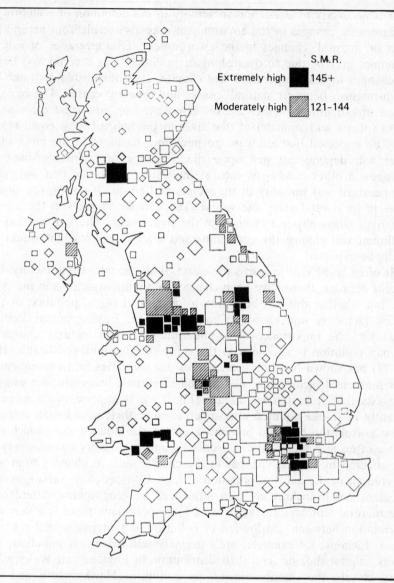

Fig. 1.1 Chronic bronchitis: distribution of mortality in the UK, males 1959–1963.
Source: Howe (1972: 241).

methaemoglobinaemia which can lead to 'blue-baby' deaths. If one concedes that pollution by its very definition implies harm then pollution can be reduced by ensuring that the substance or physical factor in question does not come into contact with susceptible targets. One cannot therefore say of pollution that a certain substance is always a pollutant and has some absolute quality of pollution. Many pollutants in other contexts or concentrations may be important materials for production. Exposure to radiation may be a hazard in one context and an instrument of medical treatment in another.

The aim of pollution control should be to reduce the harm of a potential pollutant. In formulating an economically efficient pollution control strategy it is necessary to concentrate on the damage caused by emissions rather than just their volume. Pollution consequences will depend on the one hand upon various environmental characteristics of the area (topography, river systems, climate), and on the other hand the distribution and composition of animals, crops, parks, materials, as well as the socio-economic and demographic characteristics of the population (see C. M. Wood, et al. 1974; Liu 1979).

It is clearly most important to control the emissions and their pathway to the most vulnerable targets. The most efficient pollution control strategy may not be the one that controls most emissions. For example, where pollution emissions from agricultural, mining and other non-point sources exceed those from point sources (i.e. from clearly defined and limited areas such as chimneys or drains) and urban run-off, control of the latter may still result in greater benefits as the population affected is much larger (see Watson 1978). Control of pollution should take into consideration the local use of environmental resources (rivers, land, sea, air). Thus in April 1978 the National Water Council recommended that water authorities in England and Wales should establish water quality objectives for rivers, dependent upon use, and then adjust accordingly the regulations which govern permitted emissions levels. Different standards for pollution control would depend upon the varying use of the river for public water supply, game or coarse fishing, recreation such as bathing and canoeing, industrial water abstraction, and so on.

Different pollution control strategies

One of the most important objectives of policy-making is that a problem should be tackled in the most cost-effective way possible. The most cost-effective policy is the one that costs the least in order to achieve a particular objective. If one is to determine the most cost-effective means of pollution control it is necessary to compare a wide range of possible strategies. The main strategies are as follows:

1. The initiation of a pollution event or accident can be prevented
The chance of an oil tanker accident, for example, can be lessened through strengthening shipping rules. Or the use of a persistent and polluting pesticide can be avoided by selection of plants that are more tolerant to insect damage, or by using an alternative less polluting pesticide.

The pollution problem

2. Pollutants can be diverted away from sensitive targets
Oil from a tanker accident can be dispersed and treated. In order to deal with a continuous pollution problem, the Great London sewage scheme of 1855–65 was prepared by Bazalgette. The scheme redirected the city's sewage away from the centre of London to discharge points at Beckton and Crossness, some 18 km downstream from London Bridge. The sewage was held in large reservoirs and released on the ebb tide to ensure that it would be carried out to sea (Holloway 1978). Waste disposal schemes typically involve disposal of toxic and waste materials at sites where little harm occurs.

3. The targets of pollution can be made less vulnerable
The farmer who tries to grow strawberries when there is fluorine pollution is unlikely to succeed, whereas he would more likely be successful with wheat, tomatoes or potatoes. The harm caused to buildings can be reduced by adding protective substances to paints.

4. Harmful substances can be made harmless
Bacteria in drinking water can be killed with the introduction of chlorine, waste gases can be condensed, and sewage treated. This is the form of pollution control most widely adopted as part of pollution control policy.

5. Pollution can be reduced by modifying the production process
Thus when steel production adopted the fluidized-bed combustion method of producing steel there was a dramatic reduction in water-borne residuals. Greater efficiency in production often leads to a reduction in pollution per unit of output. In the US, for example, the amount of energy needed to generate a kilowatt hour of electricity fell by just over 35 per cent between 1948 and 1968 (Maddox 1975). The use of electricity and gas instead of coal by industry and in people's homes has resulted in dramatic reductions of smoke pollution and smogs in Britain, especially during the late 1950s and 1960s. The adoption of so-called alternative technologies such as solar collectors or windmills instead of more conventional power stations may help to reduce pollution caused by energy production.

6. Changes in product design can change the demand for resources and the pollution caused in production and use.
In the 1950s a glass milk bottle in Britain weighed about 18 ounces (510 g) but by 1975 milk bottles weighing only 8 ounces (227 g) were on trial. Improvements in melting glass resulting in furnace efficiency improving at a rate of 3.5 to 4.0 per cent per annum also helped to reduce throughput of materials necessary to produce a milk bottle (*The Glass Container Industry* 1975).

7. The lifetime of products can be extended
The longer a product's lifetime the greater the reduction in flow of materials through the economy. The longer-lasting product is likely therefore to

reduce pollution from production and waste disposal. Extending the life of a car by protecting it against corrosion may consequently be an important strategy in pollution control as well as conservation.

8. Materials can be recycled

What may be regarded as a waste product or pollutant in one context is an important resource in another. In China during the early 1970s people were asked to make use of the 'four wastes' – waste materials, waste water, waste gas and waste heat. There are countless examples of the way the Chinese have converted waste products into useful industrial and agricultural inputs. Extreme care is taken over collecting and recycling glass, paper and metals while much urban waste is transported to the countryside and used for fertilizing and irrigating farmland (Orleans 1975–6; Sigurdson 1972).

9. Industrial activities can be located to prevent pollution

The location of industry north-east of Sheffield and the development of suburbs in the south-west in relation to the direction of the prevailing winds, obviously helped to reduce the impact of pollution upon the inhabitants of that city. The siting of nuclear power stations away from centres of population is another example of the deliberate attempt to reduce the possible impact of pollution.

10. Consumer behaviour can be modified so as to encourage the demand for products which involve less pollution

Recent interest in conservation may well make this strategy more important in the future. Traffic control policy favouring bus lanes and cycle ways could, for example, help to reduce the use of cars and increase the use of less polluting forms of transport.

From this listing of different strategies it is abundantly clear that a range of possibilities exists for the control of any particular pollutant. This can be illustrated simply by reference to the control of sulphur dioxide. First, the impact of sulphur dioxide can be reduced by the use of protective paints for buildings, cars and so on, or by the choice of crops in the case of agriculture. Where there is a high level of atmospheric sulphur dioxide a farmer is less likely to grow lucerne and barley than potatoes and onions. Secondly, a rise in sulphur dioxide emissions, which in Britain have increased since the 1950s due to the increased use of fossil fuels, has been partially compensated for by a fall in the sulphur content of heavy fuels. This trend is likely to increase in European countries that are supplied with low-sulphur oil from the North Sea. Thirdly, although there has been only a slight decrease in overall sulphur dioxide emissions the downward trend of mean urban sulphur dioxide concentrations at ground level has been much greater (Fig. 1.2). The reasons for this are that the pathway of sulphur dioxide has been changed by the trend towards emitting waste gases from

The pollution problem

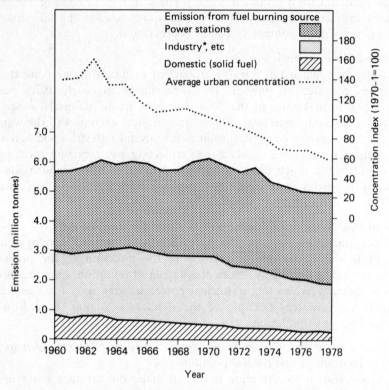

Fig. 1.2 Sulphur dioxide: trends in emissions and average urban concentrations in the UK. *Source*: Department of the Environment (1980: 3).

taller chimneys. Hence, local damage from pollution has been reduced.

A fourth way of reducing sulphur dioxide pollution is that of washing the gases before emission. At Battersea Power Station the flue gases are washed with water from the Thames. Sulphur dioxide combines with chalk in the river water to produce calcium sulphate which is then returned to the river. Recycling of sulphur dioxide to form a useful product is yet another means of pollution control. Large zinc and lead smelters are able to convert sulphur dioxide into sulphuric acid. This strategy is not, however, economically viable for the low levels of sulphur dioxide emitted in small smelters. The fluidized bed combustion techniques in power stations also reduce sulphur dioxide emissions while increasing combustion efficiency by about 10 per cent (Holdgate 1979: 79).

One of the ways of assessing which pollution strategy might be the most effective is to trace the pathway of a pollutant. Marquand (1977) has shown how pathway analysis of materials throughout the economy high-

Different pollution control strategies

Table 1.1 Path of sulphur oxides from oil (1973 quantities)

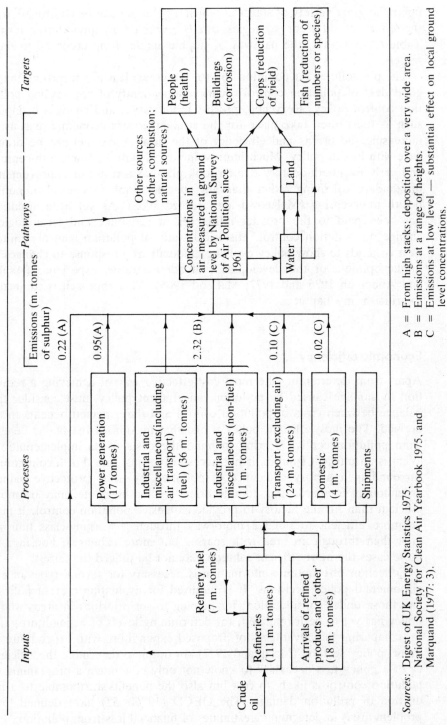

Sources: Digest of UK Energy Statistics 1975.
National Society for Clean Air Yearbook 1975. and
Marquand (1977: 3).

A = From high stacks; deposition over a very wide area.
B = Emissions at a range of heights.
C = Emissions at low level — substantial effect on local ground level concentrations.

lights the great variety of stages at which a pollutant can be controlled, and the effect of different strategies can be assessed in quantitative terms. Table 1.1 illustrates the pathway of sulphur oxides from crude oil to their targets.

The possibilities for pollution control are great. It is not surprising that a good deal of pollution control occurs independently of any explicit pollution control policy. Recycling, process modification and changes in inputs such as fuels often take place for the straightforward economic reasons of increasing the profits and efficiency of the firm. A parallel can be drawn here with health policy. Much improvement in health in nineteenth-century and early twentieth-century Britain took place as a result of improvements in standards of living rather than any specific forms of medical or public health intervention (McKeown and Record 1962). As yet little attention has been paid to the broader influences of economic and technological change on pollution control. Historical study of pollution control, limited as it is, tends to show policy evolving gradually as a response to changes in public opinion, or the development of administrative experience (Ashby and Anderson 1976 and 1977; McLeod 1965). This approach is subjected to criticism in Chapter 7.

Economic efficiency

Apart from determining the most cost-effective way of achieving a reduction in damage caused by pollution, an efficient policy must consider the balance between costs of pollution control and the expected benefits to be derived. The most efficient policy is one that maximizes the public benefit from pollution control bearing in mind the costs of its implementation. What is an efficient solution is not necessarily a just one. For a community surrounding a factory avoidance of pollution by using protective paints, greenhouses for gardening, tumble driers for washing, etc. may be more efficient than for the factory to provide expensive pollution control. It may be more efficient to build a motorway through a working-class housing area than through an area with sparser but more expensive housing. In these cases the most efficient solution may not be judged the fairest.

Collection of economic information is necessary for several types of environmental policy decisions. It is required for evaluating current policies and those under consideration, for helping to set priorities in areas where policy has yet to be formulated, for determining levels of expenditure, and for comparing the advantages of proposed expenditure with expenditure in other 'policy sectors' (Marquand 1977). In order to determine the efficiency of a policy it is necessary to know not only how much a programme of pollution control is likely to cost but also the benefits attributable to a reduction of pollution damage. The OECD (1976b: 52) have defined four problem areas in determining estimates of financial loss from pollution:

1. the specification of effects, or the problem of identifying marketable goods and services which are affected by a change in the environment;
2. the relating of effects to a specific level of environmental deterioration;
3. the problem of finding the proper prices, including the interest rate, which should be used in a monetary evaluation;
4. the problem of interpreting the calculated financial values and relating them to the monetary damage in the context of the problem under study.

Estimation of pollution damage – the dose/response relationship

The first stage in any assessment of costs or risks attributed to pollution is an assessment of actual pollution damage or the chance of damage occurring. This is generally expressed in terms of the dose/response relationship, where dose refers to the level of pollution, and response refers to the consequent damage. Saunders (1976) has reviewed studies aimed at assessing pollution damage. There are, according to Saunders, three main objectives of these studies. The first objective is to determine thresholds for total damage and for individual responses. These are usually expressed as thresholds relating to concentration of pollution and duration of exposure (Fig. 1.3). The second objective is to determine the dose/response relationship. The third objective is to measure the total damage suffered by populations, communities etc. from pollution. As Saunders points out, we are still very short of achieving these objectives, and ignorance of pollution damage is very common. This often leads to widespread disagreement over environmental effects (OECD 1976b: 53).

Ignorance of damage caused by pollution is partly a result of insufficient attention to assessment of the consequences of new chemicals used by industry. The number of chemicals currently in commercial production in the US may be as high as 70,000. Very few of these have been studied in terms of their persistence and long-term biological effects. The Toxic Substances Control Act 1976 goes some way towards remedying the problem. This Act gives the Environmental Protection Agency (EPA) the authority to demand selective testing of new and existing chemicals. Manufacturers must now give ninety days' prior notice before producing any new chemical or significantly changing the volume of production of any existing chemical. The EPA can also control the processing and marketing of any chemical substances (Council on Environmental Quality 1978).

There are practical difficulties in determining damage from pollution when this is neither self-imposed nor occurring frequently as part of economic activity. It is often necessary to use animal data as a surrogate for human data, especially in the high-dose region. Consequently dose/response

The pollution problem

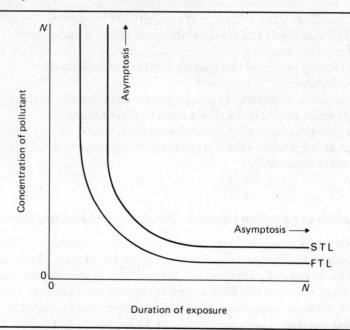

Fig. 1.3 Typical asymptotic thresholds or time/concentration curves. *FTL* = first tolerance or threshold limit of maximum combinations of *c* and *t* which do not induce a response (level of 'no effect'). *STL* = second tolerance or threshold limit of minimum combinations of *c* and *t* which induce a response. *Source*: Saunders (1976: 51).

graphs often suffer from dubious measurements of exposure. There are four types of damage surveys (Saunders 1976). First there are observational surveys of acute and episodic pollution incidents such as the *Argo Merchant* oil spill off the Massachussetts coast and the *Amoco Cadiz* oil spill off the French coast. Second, there are retrospective surveys which try to relate past pollution to damage. Third, there are studies which look at current pollution/damage relationships. These may involve surveys at work and elsewhere. Alternatively they may involve experimental studies. Finally, there are prospective studies or longitudinal studies which attempt to monitor levels of pollution and damage with the goal of determining the relationship between the two.

Lack of investigation means that there is very little information concerning some dose/response relationships such as for blood lead concentrations (Barltrop 1979). On the other hand, the dose/response relationship for radiation damage has been extensively studied (Pochin 1973). However, the dose/response relationship for most pollutants cannot be treated in isolation because pollutant damage is often related to other environmental factors (Lawther 1973). An understanding of pollution damage must take into account certain features of pollutant behaviour, such as short-term and long-term effects, persistence of the pollutant before it breaks down

Estimation of pollution damage – the dose/response relationship

through physical and biological processes, the dispersion properties of a pollutant, and interactions with other substances (Holdgate 1979). Much space could be devoted to each of these factors, but to illustrate the problems of assessing pollution damage let us consider the last factor alone. Studies of pollutants in isolation are complicated by what is called synergism and antagonism.

Synergism occurs if two or more pollutants in combination increase the level of damage above the summation of damage when in isolation. Thus when copper and cadmium occur together they are more than twice as toxic to fish than either one of them alone. The harmful effects of smoking are intensified in the presence of other air pollutants. Sulphur dioxide is much more harmful to health in the presence of particles from smoke. When rain contains sulphur and nitrogen oxides and there are also toxic metal emissions then some of the metals become soluble and are taken up by plants, so causing much greater harm than if only one pollutant had been present. One of the implications of synergism is that it may be more practicable and cost-effective to control one of the pollutants rather than both. The 1956 Clean Air Act in Britain, for example, emphasized the importance of concentrating upon controlling smoke in order to reduce the health damage caused by domestic air pollution involving smoke and sulphur dioxide. Antagonism is the opposite phenomenon to synergism: two pollutants such as sulphur dioxide and ammonia are less polluting when they occur together, as they combine to form relatively harmless ammonium sulphate.

Practical estimation of pollution damage is further complicated by a number of other exogenous factors The time factor in pollution damage is often crucial. Air pollution can cause considerable injury to the foliage of root crops, but damage to the crop is negligible if it occurs just before harvesting. In contrast, pollution just before harvesting can ruin salad crops and cause the tainting of fish. As indicated earlier, substitution of crops and protective strategies can help to reduce damage from pollution. Finally it must be remembered that damage occurs within an environmental context. The damage caused by sulphur dioxide and smoke, for example, depends to a considerable extent upon climatic factors such as temperature inversions during winter months.

Some of the problems of assessing dose/response relationships can be illustrated by reference to the controversy over low-level radiation damage. As recently as the late 1950s it was generally believed that only large doses of radioactivity were carcinogenic and that there was a threshold below which no cancers could be induced. It is only possible to demonstrate carcinogenity by comparing populations which have been irradiated with populations which have not. At low levels of irradiation large population samples are needed in order to make viable statistical comparisons. Animal studies are generally unhelpful, for the carcinogenic effect varies considerably between animal species. Consequently comparative studies have relied on just a few groups of people, the largest group being the survivors of the atomic bombs dropped on Hiroshima and Nagasaki. Then there is the group of inhabitants of the Marshall Islands in the Pacific who were exposed to radioactive fallout produced by nuclear weapons tests in the 1950s. Another group is that of people who have received several X-rays for medical purposes, and finally there are groups of workers exposed to irradiation as a result of their occupation (uranium miners and luminous-dial painters). Studies of such groups have been hampered by many elements of uncertainty. The periods and extent of dosage must often be guessed, with considerable margins of error. Exposure took place under different conditions, different age groups were exposed in different situations. Moreover, the latency period (time taken to reach the greatest possible effect) varies with different types of cancer and can be as much as twenty to twenty-five years in some instances. These factors make accurate assessment of dose/response extremely difficult.

The carcinogenic effect of small radiation doses has tended to be deduced by extrapolating downwards the dose/response curve applicable to high doses. However, given the possibilities of considerable error even of very high dose levels, there are many ways of extrapolating the curves. The International Commission on Radiological Protection (ICRP) has, in the face of these difficulties, assumed that the risk of cancer is proportional to the dose. However, records of workers at the American government's plutonium production plant at Hanford, Washington between 1944 and 1972 have suggested a much higher risk of cancer at low levels of irradiation than is assumed under a linear relationship. According to Mancuso, Stewart and Kneale the cancer risk from low levels of radiation is from ten to twenty-five times greater than commonly accepted (Barnaby 1980). However, the findings of the Hanford Study have been criticized, and there are still those who strongly believe that cancer deaths are proportional to the dose received, or in other words a linear relationship exists between dose and response (Fremlin 1980).

The lack of certainty and the difficulties of access to reliable data on radiation and hazardous materials in general allow considerable scope for interested parties either to hinder and suppress enquiry or to interpret data in their own interest. This will be explored in greater detail in Chapter 2 and in Chapter 8. Before considering the use of estimates of damage in setting standards we must distinguish between different kinds of standard.

Different types of standard

Statutory intervention to prevent pollution has taken a number of forms. It has involved legislating for various standards including ambient standards and performance standards, or the prohibition of specific activities. The standards may be laid down by the statute itself, or they may be delegated to an administrative body specializing in pollution control.

An *ambient standard* (or environmental quality standard) refers to the permissible level of pollution in the environment. Ambient water, air and noise quality standards are generally assessed in relation to damage to health, public services and amenity. Individual discharge levels are then determined so as to comply with the ambient standard. Ambient air quality standards enforceable by law have been adopted in the US, the Soviet Union, Japan and most European countries. In 1951, the Soviet Union set national air-quality standards for ten pollutants, and by 1976 the list of pollutants had been extended to 114 substances (Sharp 1976). Adopting an approach very similar to that of the US, the Japanese introduced air quality and emission standards in 1969 and created an Environmental Protection Agency in 1971. In Britain water pollution control strategies have been influenced by environmental quality objectives, but these are only used as an aid to the setting of emission standards. They have no standing in law.

The difference between *performance or emission standards* and *specification or design standards* is that the former deal with the objective to be accomplished and the latter specify how an objective must be accomplished. In Britain the 1863 Alkali Act legislated for a precise performance standard when it required that 95 per cent of the hydrochloric acid must be recovered from the emissions of alkali works. In 1874 the Act was amended to allow the Alkali Inspectorate, whose job it was to enforce the legislation, to determine the best practical means whereby a specific list of air pollutants could be controlled. This delegated responsibility in standard-setting allowed the Inspectorate to fix its own presumptive limits which were enforceable by law. While the Alkali Inspectorate (now Alkali and Clean Air Inspectorate) is responsible for standard-setting and enforcement, this joint responsibility is not always held by one administrative body. In Sweden, the National Franchise Board is concerned with permit-granting while the National Environmental Protection Board deals with supervisory and policy matters (Lutz II 1976).

Whereas the performance standard allows the polluter to find the most cost-effective way of achieving a certain emission standard, the specification standard generally requires compliance with certain design specifications. The 1956 Clean Air Act imposed restrictions upon the design both of industrial/commercial furnaces and chimneys, and of domestic hearths, boilers and chimneys. The local authority is empowered to specify the height of a chimney so as to prevent local air-pollution problems. Design standards specify design characteristics (for motor vehicles, etc.) which help to limit levels of emission. Specification and design standards are generally preferred when there are administrative difficulties, such as monitoring, in enforcing a performance standard. Performance and specification standards can either apply to the current situation or to some future date.

Both the US 1970 Clean Air Act and 1972 Federal Water Pollution Amendments set standards to be achieved within time limits.

Uniform emission standards are most appropriate when several countries discharge into a common waterway such as the Rhine or into the Mediterranean. Britain tends to favour standards based on ambient conditions, but does support the use of uniform standards for discharges from ships and oil rigs in international waters (Gardiner 1980).

Prohibition of specific activities generally occurs when less harmful alternative activities are practicable. Prohibition can be applied at a national or local level. Thus DDT has been banned in many countries when other pest control measures have become more acceptable. At a local level industrial location may be influenced by pollution considerations. Local authorities may exert an influence in this respect when planning permission is required, and pollution control administrations may prohibit industrial activity in certain areas by withholding permits or consents or by making their conditions so stringent that industry is obliged to locate itself in an area where less stringent controls are required. Given great differences in local conditions and demands upon local resources, some variation in standards and, if necessary, prohibition of polluting activity, make economic and political sense.

Standards based on threshold limits

Although our knowledge of dose/response relationships is limited and some reservations about their accuracy exist, they remain an important basis for establishing standards. The EEC's long-term policy is to establish dose/response relationships on which to base air quality objectives and eventually enforceable air quality standards (Johnson 1977). Many European countries including the UK have adopted factory air pollution standards based upon Threshold Limit Values (TLVs) set in the US by the American Conference of Governmental Industrial Hygienists (Frankel 1978). TLVs are defined as the highest 'safe' dose. Unlike standards adopted by the Soviet Union, these TLVs do take into consideration the costs of reducing risks from pollution in setting the standards. Standards in the Soviet Union are based purely upon scientific evidence of harm; as such they may be called natural standards.

The World Health Organization (WHO) (1972) has also set air quality criteria and goals for pollution control. These are commonly adopted by countries for setting national guidelines. A WHO working party has suggested two standards for air pollution involving smoke and sulphur dioxide. The first is for a standard of 250 $\mu g/m^3$ of smoke particles and 500 $\mu g/m^3$ of sulphur dioxide measured in terms of a twenty-four hour mean concentration in air. The long-term standard is for 50 $\mu g/m^3$ of smoke and 80 $\mu g/m^3$ of sulphur dioxide. These two standards represent estimated threshold levels of pollution. The first standard represents a concentration above which hospital admissions appear to increase. The second standard represents a level of pollution critical for the survival of lichens (Holdgate 1979: 149–50).

In 1970 the WHO laid down standards for nitrate concentrations in public water supplies. The concentration of nitrates recommended by the WHO is less than 50 mg/l although levels up to 100 mg/l are considered acceptable. Above this the chances of infants contracting methaemoglobinaemia are considerably increased. While there has been an increase in nitrate concentrations in some British rivers since 1945 as a result of increased disposal from sewage works and run-off from agricultural land following the application of chemical fertilizers, nitrate levels have very rarely exceeded 10 mg/l in public water supplies. Water of high nitrate concentration is usually blended with water of low nitrate concentration before distribution (Department of the Environment 1978: 32).

Where there is an obvious threshold for damage the case for basing a standard on this is strong, assuming in general terms that the costs of imposing the standard are considered less than the environmental damage or health risk associated with uncontrolled pollution. However, where there are no thresholds or where the difference between costs and benefits dictate that some damage should be allowed on the grounds of economic efficiency then it is necessary to assess and balance the benefits, costs and risks. There are four approaches commonly adopted which depend upon revealed preferences, expressed preferences, cost-benefit analysis and environmental impact analysis.

Revealed preferences

Revealed preferences are what the public seems to accept in practice as an adequate balance between risk and benefits. The public seems to accept risks from voluntary activities such as mountaineering, skiing, hunting or smoking more readily than similar risks from involuntary activities involving pollution and other hazards. In the US thirty-six times as much money is spent on saving a person from death due to radiation exposure as from an accident resulting from the way a road is constructed (Howard *et al.* 1978). The acceptable level of risk also appears to be inversely related to the number of persons exposed to that risk. The assumption behind the use of revealed preferences is that they reflect an agreed assessment of the costs associated with pollution. However, differences in opinion about the costs of pollution are reflected in the variations of court awards for damages. In the US, awards granted in noise nuisance cases brought against airport authorities have been highly variable (OECD 1976b).

The level of awards or the amount of money currently spent upon preventing pollution or accidents reflects the state of public opinion and the conflict of interests which arises in different parts of the economy. Thus the fact that 2,500 times as much money is spent on saving a life from accidents in the pharmaceutical industry as in agriculture (Sinclair 1972) can hardly be claimed to represent agreed practice. On the contrary, it is evidence that more money should be spent on preventing accidents in agriculture. The lack of a strong trade union movement in the agricultural sector partly explains the apparent acceptance of more easily preventable risks to life.

Fischoff *et al.* (1979) claimed the following limitations to the revealed preference approach. First it assumes past behaviour is a valid guide to present (and future) preferences. This is a dubious assumption in a rapidly changing world. In Japan, for example, claims for damages from mercury, arsenic and sulphur dioxide pollution have increased substantially over time (OECD 1976b). Secondly, revealed preferences are politically conservative as they reflect current economic and social arrangements. Distributional factors such as who benefits and who suffers most risks are ignored. Thirdly, it tends to underestimate damage costs which occur a long time after exposure. Finally it makes strong but often unsubstantiated claims about the rationality of decision-making. Ignorance of risks is obviously a key factor here, but so too is the whole issue of the psychology of decision-making and the way risks and benefits are assessed by the public.

Expressed preferences

Another method of evaluating pollution damage and risks of pollution is to conduct surveys into people's attitudes about pollution. Various surveys have used job evaluation techniques of ranking public priorities that need attention (The Rotheram Study 1973; Frederickson and Magnas 1972). Expressed preferences have some advantages over revealed preferences. They allow the investigator to design a survey to meet defined objectives. They also reveal current preferences rather than past preferences.

Some of the problems with this type of study are common to those of revealed preferences. The public assessment of pollution damage and risks if often inaccurate. Fischoff *et al.* (1979a) demonstrated that in the US people overestimate the death rate for several risks such as botulism, tornados and floods which arouse fear, but underestimate the death rate from chronic causes of death such as heart disease and cancer. Various other biases creep into risk assessment. Recent events such as an oil spill disaster can seriously distort public opinion. Overconfidence and desire for certainty can lead to the neglect of damage and risks which are uncertain (Slovic *et al.* 1979).

Public opinion surveys suffer from bias in the way questions are framed, and even the most sophisticated questionnaires have given rise to very unreliable results (Ashby 1976). Frequently there is a poor correlation between verbal and actual behaviour. Thus expressed preferences are often poor predictors of how people will react in a real situation. Behavioural studies also have a tendency to confuse the actual extent of a pollution problem with prejudices and preconceptions about it. Rieser (1973) examined attitudes towards noise from traffic. In a survey of traffic noise from a heavily used road he was able to show that those living in houses directly beside the road suffered more from noise than those living in spacious Edwardian villas set well back from the road. Yet when a survey was undertaken to assess attitudes towards the noise those living in the Edwardian villas appeared to be more disturbed by the noise. Both revealed preferences and expressed preferences are sometimes part of a more general approach in decision analysis called cost-benefit analysis.

Cost-benefit analysis

Cost-benefit analysis attempts to reduce a multi-dimensional problem to one of balancing monetary costs and benefits. The Roskill analysis of the proposed third airport for London attempted to reduce construction costs, noise disturbance, time savings in reaching the airport (etc.) to a single monetary value. Part of the analysis was based upon expressed preferences. People were asked 'what is the minimum sum you would accept to reconcile yourself to the increase in aircraft noise to which you are, and in the future will be, subjected?' Travel time costs were based in part upon current wage rates – the revealed value of time.

Cost-benefit analysis typically attempts to place a market value on pollution. The Beaver Committee which was set up after the London smog in 1952 tried to establish the benefits of air pollution control by estimating the total damage. This was divided into two types: direct costs and the loss of efficiency. Direct costs included damage to laundry and paintwork, corrosion of buildings, damage to various goods, and additional medical service costs. Loss of efficiency included reduced productivity of work, damage to soil, crops and animals, and interference to transport (Beaver Committee 1954).

The professed advantage of cost-benefit analysis is that it avoids all kinds of myopia, as all impacts of a project can be included in the analysis (Bohm and Henry 1979). In practice, there is a tendency to neglect impacts on which it is most difficult to put a price tag. Cost-benefit analysis which is used for road planning in Britain ignores the costs of noise and nuisance from traffic. The Ministry of Transport's analyses indicate that 80 per cent of the average benefits of a road improvement scheme result from time savings and 20 per cent of the average benefits result from accident savings (Transport Policy 1976: 99). The Beaver Committee's report also ignored the amenity costs of air pollution such as damage to the aesthetic character of monuments, reduced pleasure from recreation and the costs of suffering and bereavement. A persistent criticism of cost-benefit analysis is that it fails to include many benefits and costs related to a planning proposal (Self 1975). Difficulties in attributing economic costs and benefits and assessing the effects of time and uncertainty make cost-benefit analysis less reliable than many of its supporters would claim (Sandbach 1980).

The fact that cost-benefit analysis attempts to put a market value on costs and benefits lead to another criticism, which is similar to that directed at revealed and expressed preferences, namely that because market values reflect social and economic arrangements the use of cost-benefit analysis tends to reinforce these social and economic arrangements (Edwards 1977). One of the ways of avoiding some of the pitfalls of cost-benefit analysis is environmental impact assessment.

Environmental impact assessment

Environmental impact assessment was conceived in the US in recognition of the practical limitations of cost-benefit analysis. The National Environmental Policy Act 1969 (NEPA) required an Environmental Impact Statement (EIS) for all 'major federal actions significantly affecting the quality of the human environment'. The environmental impact statement had to contain statements on the impact of a proposed action, any unavoidable adverse impacts, alternatives to the proposed action, the relationship between local short-term uses of the environment and the maintenance of long-term productivity, and finally any irreversible or irretrievable commitment of resources. Similar legislation has been adopted to cover private actions in many American states such as California. Several European countries have also followed the American example and some form of environmental impact analysis is required by law in Germany, France and Ireland. In other countries such as Britain environmental impact analysis has been used on a few occasions but is not as yet required by law (Wathern 1976; Cairns 1978; Elkington 1980).

Environmental impact analysis depends upon numerous possible techniques for evaluating the widespread impacts of major proposed activities. Mitchell (1979) has reviewed the merits and limitations of four types of technique: check-lists, overlays, matrices and networks. Check-lists and overlays are less sophisticated but are often used during the initial stages to identify major areas of concern. Check-lists detail a range of primary impacts to be considered. Categories of impacts are usually subdivided into different areas of concern such as ecology, pollution, aesthetics and direct effect on man. The idea behind check-lists is that a broad view of all impacts will help to identify problems, and to ensure that major environmental considerations are not overlooked.

The overlay method involves the representation of the area affected by a proposal in the form of a map with the study area divided into sections based either upon a grid system or certain features such as topography and land use. Data are then collected for key characteristics of the area which are likely to be affected by the proposal. These are split into various categories such as animal and plant distribution, land use, topography, climate, geology, hydrology, soil types, physiography, demographic features, housing, water distribution. The next stage in the analysis is to estimate the positive and negative effects of the proposal on each of these categories. The estimated impacts are mapped on transparent overlays with light shadings representing low impact and dark shadings representing high impact. When this is done for each category a composite picture is formed when the overlays are superimposed. The cumulative shadings help to reveal areas of greatest impact. The problem arises, as it does in cost-benefit analysis, of making different impacts commensurable. The more refined techniques of environmental impact assessment adopt quantitative techniques which have been criticized for enhancing dubious and quasi-scientific predictions based upon questionable values attached to different types of impact (Bisset 1978).

The matrix approach identifies project actions along one axis and various environmental characteristics and conditions along the other axis. If an action has an impact upon one of the environmental characteristics or conditions then this is marked with a weighting to indicate the magnitude of the impact and another weighting to indicate the importance of the impact. In this type of analysis no attempt is made to add the numbers but they are used to indicate major areas of concern.

Network analysis is more sophisticated than check-lists, overlays and matrices in that it recognizes the chain of interactions between project activity and the environment. Air pollution from a proposed development may have a direct effect upon the plant community, but some of the effect may also be due to changes in climate resulting from the pollution – the reduction in sunlight, for example. The network analysis starts with a list of activities which would arise from a proposed project and attempts to trace the subsequent series of impacts and their interaction. Unfortunately, network analysis is plagued by lack of information about cause and effect. Consequently it is often more practical to turn to other less sophisticated techniques.

Conclusions

It has been evident that the development of pollution control policy and its implementation should be based on an appreciation of how pollution can be controlled, which approach is likely to be most cost-effective and what damage and risks are involved. Estimates of the dose/response relationship have been used as a guide for standard-setting. A broader investigation into the impact of pollution is also used in policy-making. Estimates of pub-

The pollution problem

lic preferences as well as costs and impacts of pollution have been used in this connection. However, lack of knowledge and uncertainty are difficulties frequently encountered. They are such important factors that allowance should be made for them before formulating policy.

CHAPTER 2

Pollution hazards: the problems of uncertainty and risk

Hazard management

In the previous chapter we saw the possibilities of several strategies of pollution control. The same reasoning applies to the control of accidents which cause pollution. There are essentially three possible strategies: the prevention of events which lead to a pollution incident, prevention of the pollutant reaching its target, and the mitigation of the consequences if this occurs (Fischoff et al. 1978a). Prevention of events includes not only safety precautions for operating a particular technology, but also the choice of technology in the first place. In the case of pesticides the choice of manufacturing biodegradable pesticides prevents the risks from more persistent pesticides. Likewise the choice of coal-fired power stations avoids the risks of radiation from nuclear power, but accepts the likelihood of pollution from acid rain.

The prevention of pollution can be taken back a stage further. The need for pesticides can be reduced by selecting plants with greater tolerance to insects. Similarly the need for more energy can be reduced by energy conservation. If needs cannot be modified and alternatives cannot be found then the harmful consequences of production and use must be contained – by preventing pesticide run-off from land into rivers, or by reducing radiation escape to a minimum by containment. If this aspect of hazard management fails then the next stage is to prevent exposure to the hazard. Thus foods with high pesticide residues should be avoided, and exposure to radiation could be lessened by provision of shelters. Finally, if exposure to the pollutant does occur then medical intervention might in some cases, but not as yet in the case of pesticide ingestion and radiation exposure, mitigate the consequences. This approach to hazard causation and the possible stages of intervention is illustrated with reference to pesticides in Fig. 2.1.

Each stage of hazard management may in itself trigger off other hazards.

Pollution hazards: the problems of uncertainty and risk

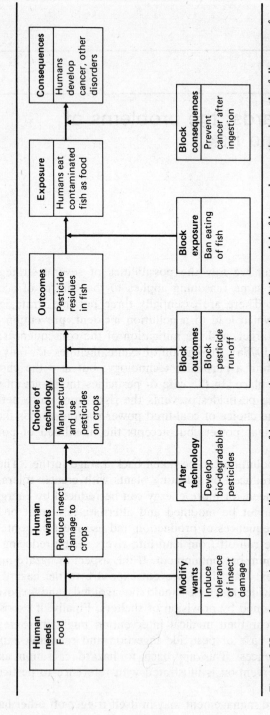

Fig. 2.1 Hazard causation and possible stages of intervention. Expansion of the model of hazard causation into the full range of stages extending from human needs to consequences. The case illustrated involves the use of pesticides to suppress crop damage. It serves as a good example of the situation in which 'down-stream' management options involving events and consequences are not very promising or even possible and 'upstream' options involving human wants and choice of technology are most likely to succeed. Source: Fischoff et al (1978a: 19).

Use of biodegradable pesticides, for example, while reducing public risk increases occupational hazards. Likewise the use of coalite instead of coal for domestic heating reduces public risks from smoke but increases occupational risks from the manufacture of coalite. Energy conservation, while reducing risks from energy production, increases risks from the manufacture of asbestos insulation board.

Hazard management consequently involves numerous possible strategies. Improvements in choosing appropriate strategies depend upon better information on risks and costs of control, upon the political balancing of different types of hazard, and upon containing interests in hazard management which use selective data, and take advantage of the public's lack of information to protect developments which are of dubious public benefit.

Uncertainties of risk

It is often presupposed that pollution control policy must be based upon known or at least predictable consequences. Borch (1968) has shown how experts who point out the uncertainties and gambles involved in decision-making are often resented. Nevertheless, if one is to come to terms with pollution hazards one has to come to terms with a large measure of uncertainty. Often much more experience is necessary before one can say with any degree of certainty what is likely to be the chance of a major incident such as a nuclear reactor core melt-down. Even then there may be a change in relevant circumstances such as threats of sabotage and working conditions. There is often an attempt to disregard uncertainties, and yet not uncommonly there is great public concern about possible but unestimatable risks that could lead to serious accidents and pollution incidents. Moreover different types of uncertainties and pollution risks evoke different public reactions. The question arises as to how one can equate the continuous, numerous and largely known risks attributed to pollution from the use of fossil fuel power stations with the more dramatic, rare and uncertain possibility of several thousand casualties from radiation damage following a civil nuclear reactor disaster. The balancing of different types of risk is often more a matter of political judgement than simple calculation, despite the common attempts to make them numerically commensurable.

In the case of an individual who smokes cigarettes, one can be fairly certain about the probable effects: the smoker of ten cigarettes a day for a year has about a 1 in 400 chance of dying from smoking (Royal Commission on Environmental Pollution 1976b). About one in four regular smokers is killed by smoking. In the US alone cigarettes cause well over 100,000 premature deaths every year (Peto 1980). Unfortunately there are many pollutants, particularly synthetic chemical products, whose effects have been either little studied or poorly investigated. If policy is to be efficient, uncertainty about risks must if possible be reduced but in any case be taken into consideration. The same applies to the uncertainties about costs of reducing risks. According to Peto (1980), estimates of direct financial costs of controlling hazards can differ by one or two orders of magnitude. Indirect costs, such as increased unemployment, loss of markets, can also be exaggerated. Uncertainty about costs of control as well as uncertainty about the number of cancers and other diseases which could be prevented by control gives enormous scope for distortion as well as genuine disagreement among pressure groups.

Even when risks are known, as in the case of cigarette smoking, industry's response has been to try and maintain its promotions and to seek new markets. In recent years, for example, there has been a massive sales drive by major tobacco companies in Third World countries. Succeeding governments have failed to levy taxes on cigarettes at a sufficiently high level to make any major contribution to preventing lung cancer. Indeed the price in real terms of a packet of cigarettes in the UK was less in 1978 than ten years previously. Peto (1980: 297), who has tried to take a balanced view of the epidemiology of cancer, still finds it necessary to point out that the response by industry

> ... has usually been to delay acceptance of the findings; to minimise their relevance to current practice, and in general to delay or obstruct any hygienic measures which will cost money. Even when human danger has been unequivocally demonstrated, industrial consortia may actively lobby for controls so weak that (as with the new UK government regulations limiting inhaled asbestos to 1 fibre/ml from 1981) they have no reasonable safety margin. Large amounts of money are available to mount press or TV publicity campaigns about the homely apple pie virtues of asbestos, journals financed by the tobacco industry run populist articles which misrepresent research results to lay readers, and some US television journalists are explicitly told always to censor all reference to the dangers of smoking.

In some pollution incidents, the type of events which could lead to an accident may not be fully grasped. For example, the possible sequence of events leading to an oil tanker accident are numerous, and every year such accidents reveal inadequacies in the design of tankers as well as rules of conduct governing tanker operations. Events leading to accidents continue to surprise us. It is necessary therefore to consider in more detail the surprising, but conceivable, accident.

The surprising, but conceivable accident

On 10 July 1976, following an explosion, a toxic cloud heavily contaminated with dioxin was released from the ICMESA chemical factory at Seveso, Italy. Trees and vegetation were heavily affected and a large number of animals died soon after the incident. By June 1977 there had been 135 confirmed cases of chloracne in eleven towns in the area and there followed a new wave of the disease in December 1977. There was also a significant number of spontaneous abortions and other clinical symptoms of health damage to the local population (Strigini and Torriani-Gorini 1977; Bonaccorsi et al. 1978).

On 28 March 1979 several water pumps stopped working in the Unit 2 nuclear power plant on Three Mile Island near Harrisburg, Pennsylvania. Within the ensuing minutes, hours and days a series of events involving mechanical, human and institutional errors led to the worst nuclear power industry accident that has occurred to date. There were no immediate fatalities but during the following year radiation damage in the neighbouring community began to reveal an increased level of miscarriages and babies born with deformed thyroid glands.

On 31 March 1980 the *Alexander Keilland*, a 'floating hotel' oil rig in the Ekofisk oil field, broke up and sank in heavy weather, claiming 123 victims.

Each of these events surprised the public and experts alike. They also helped to draw attention to the problem of how one should deal with other possible accidents and pollution events in the future. It is often claimed by economists that one should discount future risks, costs and uncertainties on the grounds that changing tastes and changing technology may alter these risks and costs (Goodin 1978). One might point to the uncertainties of future coal supplies that appeared to be likely in the UK in the 1860s. Had such risks of a future coal shortage been considered as relevant to contemporary policy then no doubt it would have been wise to conserve or ration the supplies of coal. As it was, new coal finds, better extraction technology and new energy sources such as oil and gas, changed the situation and disposed of the uncertainties and risks which had appeared in the 1860s. However, the uncertainties and risks connected with nuclear power are of a different kind. What might happen in the future might equally well happen tomorrow or even today. The risk of a major nuclear disaster in the next thirty years cannot simply be discounted as a problem that might be overcome by future generations.

When the events which could lead to a pollution accident have not occurred the possibilities can be analysed in a hypothetical way, using what is termed an 'event-tree' or a 'fault-tree'. Event-tree analysis considers a variety of events which could initiate an accident. An assessment is then made of the possible consequences of each event. In the case of a nuclear power accident the initiating event may be a pump failure. The fault-tree approach, on the other hand, starts with the faults that result in an accident and tries to identify the conditions which allow the fault to take place. Identification of potential events and faults obviously helps in taking precautionary measures, but can also be used to estimate risks of accidents.

Pollution hazards: the problems of uncertainty and risk

Nuclear power

In the Rasmussen report (Atomic Energy Commission 1975) 'event-tree' and 'fault-tree' analyses were used to assess the combination of faults and events which could lead to a nuclear power accident. On this basis, an attempt was made to calculate the probability of a 'melt-down'. A 'melt-down', involving perhaps 3,000 to 4,000 or more immediate deaths and many more long-term radiation-induced illnesses and deaths (Table 2.1), would follow from failures in the coolant system which allow the fuel rods to overheat and form a mass of molten and highly radioactive material. This might then burn its way downwards through the floor of the reactor – a possibility the Americans have labelled as the 'China Syndrome', as China would be the general direction in which molten mass travels. The chance of a reactor core melt-down was estimated to be about 1 in 200 million years of reactor operation.

While 'event-tree' and 'fault-tree' analyses can help experts to judge pathways to disaster they may also lead to over-confidence about ways in which technical failures can come about (Slovic et al. 1979). Many accidents have been caused by chains of events which have not been anticipated. The Rasmussen report has been criticized for ignoring the possibility of unanticipated events and for this reason is likely to underestimate the risks (Sørensen, 1979). Rasmussen himself is reported to have been surprised by the series of failures which led to the Three Mile Island accident. In the light of such uncertainties, a Ford Foundation study group's report *Nuclear Power: issues and choices*, which suggests a one in four chance of a major nuclear accident involving a melt-down before the year 2,000, may offer a more realistic assessment of the risk (Bugler 1979).

The Kemeny Report was commissioned by the President to investigate the accident at Three Mile Island. It identified a lack of expectations of the type of technical failures that occurred as a key feature of the accident. The US nuclear industry and the Nuclear Regulatory Commission had trained its operators to deal with the so-called large-break accident, but less preparation was made to deal with the series of minor events which could lead to a major accident. In these circumstances inappropriate oper-

Table 2.1 Rasmussen's estimate of possible consequences of 'An Extremely Serious Accident'

Prompt fatalities	3.300		
Early illness	45,000		
Thyroid nodules	240,000	over	30 years
Latent cancer fatalities	45,000	over	30 years
Genetic defects	30,000	over	150 years
Economic loss due to contamination $14 b.			
Decontamination area 8,300 km^2			

Source: Bugler (1979: 31).

ator action resulted in overheating of the core and extensive fuel melting. The Rasmussen Report had not anticipated events which led to the formation of a non-condensable gas bubble in the reactor vessel. It was this event which confounded the technicians at Harrisburg.

Etemad, former project leader in the French nuclear industry, has claimed that it was only because the nuclear reactor was relatively new that a major catastrophe was averted: 'if the fuel had been more highly irradiated – say after the third partial refuelling – the decay heat of the reactor throughout the emergency sequence would have been about doubled. In that situation, and with the known circumstances and technical ability to get rid of waste heat, substantial overheating and a major catastrophe could not have been avoided' (*The Guardian* 17.1.1980).

Another example of an unexpected nuclear power accident occurred on 22 March 1976 at the Brown Ferry nuclear power plant in Alabama. An electrician, in violation of standard operating procedures, was using a candle to check for an airflow through a wall and accidentally ignited some foam padding around the cable tray. A fire spread rapidly knocking out all five emergency core-cooling systems of the No. 1 reactor. In Britain the worst accident occurred at Windscale on 8 October 1957. A fire broke out in the No. 1 reactor core due to an error by the physicist in charge. An attempt was made to put the fire out using liquid carbon dioxide. After this failed the local fire brigade was called in, and the fire was eventually put out. A large amount of radiation was released, but most of it was trapped by filters in the stacks. It was these filters known as 'Cockcroft's Folly' that had been put in at the request of Cockcroft despite some opposition from colleagues, which prevented an accident turning into a major catastrophe. As it was the government had to throw milk away from an area of more than 1,300 km^2 as a result of contamination by iodine 131, the hazardous radioisotope responsible for causing thyroid cancers (Bugler 1979).

The relatively good safety record of the nuclear power industry and the few 'surprising' accidents that have occurred do not detract from the large number of possible accidents that could occur during every stage of the nuclear fuel cycle, including mining, enrichment, transportation, power generation, reprocessing and waste disposal. For example, Kinnersley (1980) claims that transportation of 'spent fuel' by rail from Dungeness and Sizewell through London to Windscale could lead to a major catastrophe. If only 10 per cent of the radioactive material was released, a fan-shaped area of 8 km^2 would have to be evacuated for about five years. If all the nuclear wastes were spilt then an area with a radius of 19 km, the major part of central London, would have to be evacuated.

This type of accident makes nuclear power very vulnerable to sabotage and terrorist action, clear possibilities in a period of growing political instability. The Royal Commission on Environmental Pollution noted that plutonium, which is a by-product in thermal reactors and a fuel in fast-breeder reactors, 'appears to offer unique potential for threat and blackmail against society because of its great radioactivity and its fissile properties.' (1976b: 202). To demonstrate just how vulnerable nuclear power is, in June 1975 a member of the West German Social Democratic Party walked through the security barriers of the world's largest operating nuclear power plant with a two-foot long bazooka, capable of blowing a four-inch hole in the plant's pressure vessel, and presented it to the director of the plant (*Rights* 1976 **1** (2) 3). The following types of subversive risks associated with nuclear power are realistic possibilities:

1. sabotage inflicted on a nuclear installation or on the transportation of nuclear material;
2. the detonation of a nuclear bomb constructed with fissile material obtained from, or during transit between nuclear installations, and
3. the dispersion of fission materials as a carcinogen.

Apart from risks associated with subversive human intervention there are serious technical problems which could lead to disaster. There is a long-term problem of storing highly active nuclear wastes. The wastes must be stored for about 1,000 years until the radioactive isotopes have decayed to an insignificant level. In Britain these wastes are stored at Windscale and to a lesser extent at Dounreay, the site of a prototype fast-breeder reactor. At present they are stored as acid solutions in specially designed storage tanks which need to be cooled, but the long period for which these wastes must be stored requires a more permanent means of storage. As yet there is no such satisfactory method of disposing of these dangerous substances, although the intention is to convert the wastes into a glass-like material which requires substantially less maintenance. The Royal Commission on Environmental Pollution (1976b) believed that an acceptable solution would be found to the storage problem, but until then Britain should not be committed to a large extension of the nuclear fission power programme.

During the winter of 1957–8 there was a major nuclear disaster in the Soviet Union. Only recently has attention in the West been drawn to the incident by Medvedev, the Soviet dissident. He claimed that hundreds were killed and a vast area was closed to the public. Evidence suggests that the accident arose following the failure of a cooling system on a high-level waste-storage tank. Two lakes were contaminated. Since the accident more than thirty names of small communities of less than 20,000 people have been deleted from maps of the area (*New Scientist* 10 Jan. 1980). As yet no such major incident has been reported elsewhere with the possible exception of a nuclear accident reported as an earthquake in 1976 at the Russian naval base of Paldiski where nuclear silos were stored (Fernie 1980). It seems only a matter of time before a major nuclear disaster takes place elsewhere.

Balancing risks

One of the major dilemmas of our time is how to make decisions about dealing with the unknowable and yet conceivable environmental risks. One possible solution is to discount uncertainty and attempt to squeeze known risks into a traditional risk/benefit framework. Inhaber (1978) tried to compare the risks of ten different energy technologies. The only fair way to assess the risks, he argued, is to make an objective assessment of the risk per unit of energy. By doing this he produced a very different assessment from that based on an intuitive comparison of the cheerful-looking solar panel perched on a roof contrasted with the closely guarded and rather awesome picture of a nuclear power plant. Furthermore, he continued, an assessment must be made of the total energy cycle and not just an isolated part of it. Risks are involved at all stages of mining, and in the manufacture of energy-producing plant and equipment. According to this type of assessment, Inhaber claimed that the risks associated with nuclear power are a little greater than those from natural gas, but considerably less than from coal, oil and the unconventional wind, methanol, solar and ocean thermal technologies.

The main reasons for the somewhat surprising results are that risks from the non-conventional systems are associated with the larger amounts of materials and labour they require per unit of energy output. The main risks to these sources of energy come not from the operation of a solar collector or windmill but from obtaining materials necessary for construction, including the mining of coal, iron and other raw materials, as well as processing them into steel, copper, glass and so on. However, the validity of the study rests partly on the important assumption that reliable evidence exists from which the rates of industrial accidents, disease and death, and the raw material requirements for industrial processes, can be ascertained. While Inhaber admits that none of these data are known absolutely, he does believe there is sufficient reliable information to make a valid general comparison.

Criticisms of Inhaber's study are of two kinds. The first type accepts the validity of the methodology but points to mistakes in calculations or to what has been left out. For example, Inhaber assesses the risk of energy produced, but not of energy consumed. Hence the risks associated with the distribution of electricity by transmission lines are ignored. Musgrove (1978) has claimed that Inhaber has over-calculated the risks associated with windmills by a factor of 100. There are also several methodological problems. As referred to earlier, there is the difficulty of making future uncertainties about low probabilities of very serious disasters commensurable with the much better-known risks of other technologies. Herbert et al. (1979) pointed out that Inhaber's study does not distinguish between catastrophic and non-catastrophic events. There is an assumption that they can be made commensurable. However, there is considerable evidence that people consider catastrophic events more seriously than non-catastrophic events even when the same number of people are affected (Fischoff et al. 1978b). As a consequence people are prepared to make greater efforts or spend more to avoid catastrophic events (Slovic et al. 1979).

The risks of a major nuclear power disaster would also be to the local community, involuntary victims of pollution. Most of the risks associated with other forms of energy production principally apply to those involved in the industry and can be considered, so long as they are known, more in terms of voluntary risks: if the risks of work are unacceptable it is usually possible in theory, if not always in practice, to look for alternative employment. This distinction between voluntary and involuntary risks is important, for it is generally accepted that people are more willing to accept voluntary risks than involuntary risks. One might argue that under normal operation of nuclear power plants the risks of pollution to the neighbouring community are minimal. Over half of the radiation to which we are exposed comes from natural radioactivity. Of next greatest importance is the contribution from medical and dental X-rays. Uranium mining and processing operations lead to a little radiation and nuclear power plants, barring accident, produce relatively insignificant amounts of radiation (Table 2.2). However, it is the unforeseen leakage or accident involving technical failures and human error or deliberate sabotage which could affect a large number of people outside the nuclear power industry that leads to fear and concern about the continued development of nuclear power (Lerch 1980).

The distinction between types of risk is important for another reason. Risks of major disasters which may be brought about by saboteurs, terrorists or anarchists necessitate stringent security measures, involving screening the workforce and the surveillance of any potential saboteurs. In the case of nuclear power its vulnerability to potential disasters requires surveillance measures which threaten democratic freedoms and civil liberties (Flood and Grove-White 1976). Benn (1979), when Energy Minister for the British Labour government, claimed that security involved an armed constabulary at nuclear plants, phone tapping, collection of personal information and dossiers, and clandestine operations which made public accountability extremely difficult.

The different types of risk are also important in the event of a major catastrophe occurring. Natural hazards, such as hurricanes and earthquakes involving thousands of deaths, may have a profound and lasting effect but nothing in comparison to a nuclear power accident involving a similar number of people. Bugler (1979: 32) commented:

> Think of the litigation after a DC-10 air crash. Then imagine what would happen after an accident of 3,000–4,000 immediate deaths – and tens of thousands of delayed cancer deaths, spread over 30 years. And

then perhaps genetically damaged children being born over the following 150 years. The great probability is that there would be a furore of huge dimensions, conceivably even the collapse of a civil government, and perhaps a backlash against nuclear power that would leave society with a choice of an immediate electricity shortage (by shutting down reactors) or force of arms (to keep the reactors working in the face of civil opposition).

These different kinds of risk and associated uncertainties make it impossible to produce any simple formula for comparing natural hazards and pollution hazards from different technologies. Nevertheless identification of risks and uncertainties should lead to more informed decision-making so long as one is aware of the dangers of reducing all elements to a common calculation of risk. The process of identification is, however, subject to social and economic influences. The way in which information on pollution is generated and bias mobilized is of crucial importance in the control or lack of control of pollutants.

Table 2.2 Dose rates in the UK from ionizing radiation

	Bone marrow	Reproductive cells: genetically significant dose (GSD)
	mrem/yr	*mrem/yr*
Naturally occurring:		
From cosmic rays	33	33
From soil and airborne	44	44
Within the body (mainly potassium-40)	24	28
	101	105
Manmade:		
Medical, diagnostic X-rays	32	14
Medical, radiotherapy	12	5
Medical, radioisotope use	2	0.2
	46	19
Fallout from bomb tests	6	4
Occupational doses (other than from nuclear power)	0.4	0.3
Nuclear power industry	0.25	0.2
Miscellaneous (mainly occupational doses)	0.3	0.3
	7	5
	154	129

Source: Royal Commission on Environmental Pollution (1976b: 16).

Pollution hazards: the problems of uncertainty and risk

The politics of uncertainty and the monopoly of information

Nuclear power generation has considerable elements of uncertainty and risks which cannot be discounted in policy-making. However, where uncertainty exists, the scope for scientific controversy and disagreement is great. Nelkin (1975) has argued that such conflict among experts reduces their political impact. Nevertheless, there is often a virtual monopoly of expertise and control of information within industry which makes it better equipped to mobilize bias where uncertainty exists.

At the Windscale inquiry into the British Nuclear Fuels' application to extend its reprocessing activities, there was profound disagreement between experts on how reprocessing might influence future energy policy, nuclear proliferation, pollution and terrorism. While the Atomic Energy Authority played down the health risks of low-level radiation, Stewart, a world renowned epidemiologist, suggested that risks from low-level radiation might be up to twenty times greater than current estimates (Chapter 1). However, uncertainty still existed as the data base on which estimates could be made was incomplete and the risks to workers in the industry would not be known until the 1980s (Fernie 1980). In this situation of uncertainty, the nuclear power industry's near monopoly of expertise placed it in a better position to legitimate its expansion and to play down the risks of accidents and pollution. Although Mr Justice Parker, the inquiry inspector, could not resolve the debates on risk of nuclear proliferation, the threat to civil liberties and the morality of leaving the waste disposal problems to future generations, he recommended that British Nuclear Fuels should be allowed to proceed without delay. The recommendation was duly accepted by Parliament.

Epstein (1979) has studied the control of information on benefits and risks of chemical products. He claims that the majority of data come from industrial resources, and this allows for considerable manipulation and suppression of data. As a consequence, uncertainty of risks is either perpetuated or concealed, and the certain knowledge of risks is often delayed. There is inevitably a tendency for industry to support research programmes that legitimise its activities. Epstein argues that cancer testing often results in inadequate or misleading data. For example, 'a 1969 review of seventeen industrially sponsored studies on the carcinogenicity of DDT by the Carcinogenicity Panel of the Mrak Commission on Pesticides concluded that fourteen of these studies were so inherently defective as to preclude any possible determination of carcinogenicity' (1979: 264). Even more seriously, Epstein gives evidence of fraudulent manipulation of data.

Industry is also in a powerful position to influence economic assessments of the cost of controlling risks, an important feature of risk analysis. Although consulting firms may be called in to make such assessments, their findings are based on information provided by industry, and so are likely to favour industrial interests. Epstein illustrated this by reference to economic impact analyses in 1974 of anticipated costs of meeting the proposed vinylchloride standard. The two consulting firms which undertook the work confirmed industry's claim that the standard would be impractical and too much of an economic burden on industry. The economic claims of costs and unemployment turned out to be exaggerated. The study failed to include benefits from greater recovery of vinylchloride as well as major costs to society of its carcinogenic properties and other diseases associated with the chemical.

Many chemicals now widely used in industrial processes have not been shown to be harmful to health until years after they have been introduced. By this time the chemicals involved may well have become of great economic value. This was the case with acrylonitrile, a synthetic polymer widely used in the textile industry. It was introduced in the 1950s and became the basis for a new industry, but it was not until 1977 that scientists at E. I. du Pont de Nemours and Company found that traces of acrylonitrile in the air were the probable causes of high rates of cancer of the colon and lung among its textile workers (Behr 1978). Delays in detecting risks of this kind and doing something about them have led to tragic cases of ill-health and disease. Unhappily, delays in understanding the dangers involved are often related to the lack of concern for workers and the public by industries constrained to maximize profits and ensure steady growth in capital. This assertion will be supported in detail in Chapter 8 when the historical development of knowledge about asbestos hazards is reviewed.

Conclusions

Different types of pollution give rise to different social responses. Consequently it is not always desirable to make a quantitative comparison of risks. Difficulties of predicting events which could lead to accidents ought not to be, but frequently are, neglected in policy. Consequently, risks from some technologies, such as nuclear power, are probably much underestimated, especially by their promoters. Uncertainty about risks and damage of pollution is often manipulated by industry as a way of avoiding extra costs of control.

CHAPTER 3

Prevention rather than cure: an alternative strategy

There is little doubt that in many instances it is profitable for private industry to seek conservation policies which reduce waste and pollution. Yet the emphasis of most environmental legislation, pollution control management, and academic treatises is to regard pollution control as a cost to the firm (see Ch. 6 for the traditional debate). Royston (1979), in contrast, has argued that too much emphasis in the past has been placed on controlling pollutants by converting them into something less harmful but useless. Instead, greater effort should be made to reduce the quantity of waste by increasing process efficiency and recycling. In this chapter we will look at some of the possibilities of and constraints on preventing pollution while maintaining other economic and social objectives.

Royston has documented many instances of what he calls 'pollution prevention pays'. This catch-phrase was first coined by the 3M Company which adopted it as a guideline for a major investment programme. While looking for a reasonable return on investment, the programme also sought to reduce pollution through product reformulation, process modification, redesigning of equipment and the recovery of waste material for re-use. Royston cites many other companies which have successfully combined a strategy of preventing pollution with corporate expansion. For example, the PLM Company in Sweden specializes in packaging materials, glass bottles and tin cans. In response to concern about environmental problems in the late 1960s it developed interests in recovery and recycling of materials. Ten years later this new side of the business accounted for half the company's turnover.

Royston argues that it is mainly as a result of tunnel vision that industry often regards prevention of pollution as an economic burden, and that the scope for productive investment in pollution control is much greater than generally recognized. However, he does recognize the existence of a number of economic, administrative and structural barriers that prevent the development of conservation and 'non-waste technology'. Consequently, initiatives to encourage pollution prevention are necessary from government, industry and the local community. This can be illustrated by assessing the potential of recycling and the constraints on the reclamation industry in Britain.

Recycling

Recycling is a very general term meaning the re-use of materials. It is worth distinguishing between three main modes of recycling: first, direct recycling to produce similar goods such as re-using lead from lead batteries to make more lead batteries; secondly, indirect recycling to produce different or inferior products – for example, recycling of mixed paper to form cardboard; and thirdly, the re-use of products such as milk bottles. Recycling can provide cheaper products, prevent disposal problems, conserve resources and reduce pollution from processing and manufacture. Recovery and re-use of waste paper allows the manufacture of products which would be considerably more expensive if produced from virgin wood fibre. It helps to conserve forests and prevents a waste paper disposal problem. Moreover less pollution and energy resources are involved in reformulating paper than in producing paper from forests. Recycling of paper and many other materials in Britain also reduces the demand for imports and consequently helps the balance of payments position.

Conservation of materials is an important consideration in recycling. If half our resources were recycled then the lifetime of reserves would be increased by a factor of 2. If 90 per cent of resources were recycled then the lifetime of reserves would be increased by a factor of 10. Moreover recycling usually indirectly involves the conservation of energy and other resources. Typically aluminium requires some 91,000 kWht of energy per tonne when produced from bauxite, but only 3,000 kWht of energy per tonne when produced from scrap. Steel produced from scrap requires about 30 per cent of the energy necessary to produce steel from iron ore. There are then several short-term and long-term interests in recycling, and it is consequently not surprising to find that a considerable amount of recycling does occur in the Western world (Table 3.1).

The scope for making better use of domestic refuse is considerable. In Britain between 1935 and 1970 the density of household waste has halved as the proportion of packaging in our consumer society has grown. Fine dust and cinder which made up the greater part of the composition of domestic waste in 1935 (57 per cent) declined to only 19 per cent of the total in 1970. During the same period the paper content increased from 14 to 33 per cent. While most of the waste (80 per cent) is still disposed of by controlled tipping, the remainder (20 per cent and increasing) is disposed of by incineration. With increasing shortages of suitable tipping sites for major urban areas, incineration has become more popular as the volume of

Prevention rather than cure: an alternative strategy

Table 3.1 Western-world Recycling Ratios

	Aluminium		Zinc		Lead		Copper	
	1963	1973	1963	1973	1963	1973	1963	1973
France	20.7	20.0	35.7	27.0	19.8	23.0	34.8	26.9
W. Germany	26.6	25.6	8.8	12.8	14.2	18.0	37.9	27.0
Italy	6.4	24.0	35.7	29.1	3.9	19.7	28.5	29.4
Japan	24.9	28.0	24.5	12.1	32.2	22.1	42.0	35.5
UK	32.4	28.5	24.7	25.0	52.1	61.6	36.3	37.9
USA	22.2	23.8	24.9	23.4	41.3	48.1	43.3	44.7

Source: Pearce 1976c.

waste can be reduced by about 90 per cent. Only a few incinerators make use of the heat for raising steam to produce electricity or for providing heating for nearby houses and factories. Moreover pyrolysis, which could produce commercial gas while decomposing the waste in the absence of air, has been commercially neglected. The US, in common with Britain, has converted very little (less than 1 per cent) of its municipal wastes to energy. In comparison by 1977 Denmark was converting 60 per cent of its wastes to energy, Switzerland 40 per cent, and the Netherlands and Sweden each 30 per cent (Council on Environmental Quality 1979: 261).

Little effort has been made by local authorities to separate and recover materials from domestic waste, with a few isolated exceptions such as the separate collection of waste paper and the extraction of metal cans by using magnets. The Warren Springs Laboratory has pioneered the development of some refuse-sorting equipment on a small budget, but on the whole extraction of materials from domestic waste has been virtually untapped and still depends in the main upon pre-industrial use of labour for separating and collecting. The recently-built pilot refuse sorting plant at Doncaster represents an initial but belated effort to further recycling in this field (Fernie 1980; Fergusson 1974).

The case of waste-paper recycling illustrates the importance domestic waste could play in expanding the profitability of reclamation and the paper industry. In 1976 about 58 per cent of waste paper came from printers, publishers, large stores and offices. This is a fairly stable source with little scope for expansion. Thirty-two per cent of waste paper came from other commercial sources and only 10 per cent came from domestic refuse. It is this latter source which provides the greatest potential for expansion. Thomas (1977) claims that shortage of wood pulp has put and will continue to put the UK paper industry at a huge competitive disadvantage. British paper producers saw their share of the home market fall from 77 per cent in 1970 to 53 per cent in 1976, and since 1965, thirty mills had to close with employment declining by a third.

Thomas claims that the only way the paper industry can hope to remain competitive is to increase its use of waste paper, but industrial myopia and timid government were blamed for the absurd situation in 1976 of having to import about 100,000 tonnes of waste paper from abroad, especially Germany, while over six million tonnes of home-produced paper were uncollected and wasted. During the same year local authority collections fell from 450,000 tonnes to 200,000 tonnes (*Materials Reclamation Weekly* 15 Jan. 1977 and 14 May 1977). Studies in America suggest that between 25 and 35 per cent of domestic waste paper could be collected. Accordingly, Pearce (1976b) believes that 1.7 million tonnes of waste paper could be collected each year.

Economic and organizational constraints on recycling

The constraints on improving recycling ratios are numerous. There are some obvious physical limitations. In the case of paper some such as cigarette papers and toilet paper are destroyed in use; some such as books, wallpaper, and building board go out of circulation; and some, because they are combined with other materials, make recycling impracticable.

Many collecting and sorting schemes have had to be discontinued because of high cost. The Oxfam Wastesavers Scheme in Huddersfield was discontinued in 1977 because of the costs of processing glass, tin and plastics. The costs of collection are particularly important in the case of glass recycling as the raw materials for manufacture – sand, soda and limestone – are widely and cheaply available in Britain (Fernie 1980). Economic interests of manufacturers using raw materials are a further constraint on recycling. Price factors play a particularly important part as recycling usually demands high labour and low energy inputs. So recycling is more likely to be cost effective if labour costs are low and fuel prices are high. The costs of waste collection, and the viability of recycling depend in part upon the size and geographical character of the collection area.

Rapid market fluctuations in raw materials and waste products have created difficulties for the reclamation industry. Between summer 1974 and spring 1975 waste-paper prices in the UK fell by 75 per cent (Waste Management Advisory Council 1976), and some waste paper was even turned away by the paper companies. Without supply and price guarantees the collectors of waste are put in a very vulnerable position. Many local authorities have for this reason been discouraged from collecting waste paper. Pearce (1976a) has shown how cycles of business in the waste-paper industry mirror almost exactly the cycles in economic activity of the British economy. However, the amplitude, or size of fluctuation, of the waste-paper cycle is much larger, involving greater booms and slumps. Consequently local authorities and charities who enter the collection business during a boom time often find themselves in difficulties when demand slackens during a slump period. To some extent market fluctuations would be lessened by increasing the capacity for processing waste paper, and in 1977 the Labour government introduced a £23 m. grant scheme to assist paper manufacturers to increase their capacity (*Materials Reclamation Weekly* 25 June 1977).

In the US the paper products made from recycled fibre have declined from 30 per cent of the total waste in 1950 to 22 per cent in 1977. Most paper producers favour using virgin pulp with prices and demand for waste paper only rising dramatically when virgin pulp has been in short supply. The potential for using more recycled paper clearly exists. For example, only about 12 per cent of US newsprint is made from recycled papers, and the amount of paper recycled is limited more by a lack of paper industry demand than by insufficient supply (Council on Environmental Quality 1979: 272).

There are several organizational problems relating to collection, sorting and separation of materials. Often this is not helped by the packaging and presentation of products. Standardization of the size and shape of returnable bottles for instance, or the design of metal cans, could help. However, this is often resisted by marketing agencies who believe that their own distinctive shape, size and colour of bottles and cans helps to sell their products. So the problem is that product specification may help the reclamation industry but not the manufacturer of the product. Product specification may lead to an overall economic advantage, but divided responsibility between manufacture and reclamation prevents an automatic response to market considerations. It is in this situation that government could play an important part in improving economic efficiency while at the same time preventing pollution (*War on Waste* 1974).

The reclamation industry is notably fragmented and has suffered from lack of research and development. This applies particularly to the problems of collection, sorting and preparation, but there are also some technical problems preventing greater efficiency in utilizing waste products. In the paper industry recycled paper products could be enhanced if there were improvements in preparing waste-paper pulps. Thomas (1977) sees consumer resistance to inferior paper products, promoted in part by advertising, as another major barrier to further waste-paper reclamation.

Product substitution, product use and the social organization of production – the case of China

Individual needs can be met in numerous ways involving the possibilities of product substitution and different ways of using the same product. A few generalizations may be helpful at this point. All things being equal, the product, such as a car, with a longer life expectancy, will demand less resources, and lessen the problem of waste disposal. The car that is fully utilized, carrying an average of four passengers, will cause less pollution and use less resources than four cars with only one passenger each. Co-operative and communal use of products may be more efficient and involve less pollution. The late E. F. Schumacher, in his many talks about the waste of resources in Western society, used to refer to the sight on a motorway of a lorry travelling in one direction carrying biscuits from the north to the south and in the other direction a similar lorry carrying not dissimilar biscuits. The message of the tale being that geographical self-sufficiency of products prevents wastage of resources and pollution involved in transportation and marketing at a distance. Prevention depends therefore upon the social organization of the economy.

In China development policy has emphasized the prevention of social problems such as disease and pollution. The development of co-operatives and communes in the 1950s enhanced the possibilities of self-sufficiency in rural areas and has facilitated the development of mass campaigns, often during slack periods of agriculture, to clean up the environment and prevent pollution. The commune structure has led to efficient use of scarce resources and products, and has also reduced the pollution problem of interregional transportation. The promotion of social objectives in economic planning has resulted in an integration of environmental policy with health, agricultural and industrial policy. Just as the 'pollution prevention pays' policy has been successfully adopted by some Western firms so too have the Chinese been successful in integrating pollution control with other social and economic objectives. Prevention of sewage pollution in rural areas has a dual role of preventing communicable water-borne diseases such as schistosomiasis, and at the same time providing a valuable fertilizer for agriculture (Sandbach 1977d; Horn 1969).

Control of industrial pollution in China has played a role in both health and economic policy. Scarcity of resources for industry has led to a comprehensive policy of multi-purpose use. For example, slag and solid waste from coke ovens, steel mills and smelters are used for making bricks. Recycling and use of waste products became part of mass campaigns in 1970 when the Chinese were called upon to make the maximum use of the three wastes – waste material, waste gasses and waste water. As a result valuable resources were recovered and pollution was reduced. Waste organic materials are widely used in the production of biogas (methane), especially in areas poor in traditional fuels. Some 7.2 million biogas digesters were constructed in China during the 1970s. This is about 100 per cent more than in India which is the only other country to have built them on a large scale (Van Buren 1980).

Finally, the widespread adoption of small-scale technology in rural areas as part of the 'walking on two legs' policy, promoted in particular first during the Great Leap Forward (1958–9) and then during the Cultural Revolution (1966–9), was an attempt to promote economic growth and self-sufficiency in rural areas. The dual role of centralized large-scale industry and the small-scale rural technology has played an important part in preventing pollution problems associated with urban sprawl and the concentration of industry in urban areas (Sandbach 1980). This is not to say that

China is free from pollution problems. Some characteristics of its economic development are producing severe health risks. In particular, China's dependence upon coal for power generation and domestic heating has led to an increase in lung diseases (Joyce 1980).

Alternative technology

The Chinese example of promoting small-scale technology in rural areas has its counterpart in the West in the form of the alternative technology movement. Clarke (1973) has made a useful distinction between four different responses to pollution problems. The first response was based on price or utility. Pollution, accordingly, is often regarded as the price we have to pay for progress. The second response is the technical fix which is to cure pollution problems and convert them into something harmless. The third response is to do away with technology. The fourth, and favoured, response was to develop technology which was non-polluting. Clarke (1973: 259) claimed 'that not all technologies are intrinsically polluting and that new forms of technology can and should be devised to remedy a deteriorating situation. Thus instead of burning fossil or nuclear fuels, with their particulate and thermal pollution, we should develop technologies such as the use of solar and wind power which are intrinsically non-polluting.'

Clarke dismissed the third anti-technology option, and so his main comparison was between the hard technology path implied by the first two options and the soft or alternative technology path of development. Alternative technologies according to Clarke should be non-polluting, cheap and labour-intensive, involve few non-renewable resources, be incapable of being misused, be understandable by all, aid decentralization, and be non-alienating (Clarke 1972).

These distinctions between hard and soft technologies were applied to many areas of economic activity such as agriculture (chemical fertilizers and pesticides *versus* organic farming), housing (energy wasting houses *versus* autonomous houses, domes and greenhouses), medicine (Western drugs *versus* homeopathy and natural medicine), transport (cars and planes *versus* bicycles, canal transport and airships), and energy (coal, oil and nuclear power *versus* hydro-electric, solar wind, geothermal and tidal power).

A common feature of the alternative technology movement in the early 1970s was the attempt to find self-sufficiency and ecological harmony within a very small, and mainly unstable, social group usually involving a single house and small co-operative. This was illustrated in the visions of Clifford Harper regularly found in the *Undercurrents* (Fig. 3.1), and was put into practice in communities like BRAD (Biotechnic Research and Development) in Wales. The tools of alternative technology have been described in

Fig. 3.1 The self-sufficient home. 1. Solar radiation trap heats water; 2. Collection of wind and rainwater; 3. Purification and storage of water; 4. Methane gas from decomposition of wastes; 5. Reservoir, fish culture; 6. Intensively cultivated vegetable garden; 7. Animals. *Source*: Harper (1973: 296).

detail by Dickson (1974) and Harper *et al*. (1976). A wide range of such alternatives as windmills, solar cells and panels, methane digesters, water wheels, and an energy-saving house are exhibited at the National Centre for Alternative Technology, Machynlleth, Powys, Wales (Hanlon 1977). Most of the alternative communities such as BRAD have only been sporadic and short-lived. The Centre for Alternative Technology has been a more permanent feature receiving considerable commercial support.

The practical application of alternative technology was largely in the hands of scientists, technologists and others who had rejected the technological imperatives implied by the 'hard technology system'. The utopian communities that sprang up were strongly guided by the belief that development of alternative technologies would facilitate more harmonious social relationships and relationships with the environment. Most of these experimental communities made use of commercially available equipment. Research and development of alternative technologies by those promoting the social and environmental advantages was rare. An important exception was the development of the New Alchemy Institute which was established in 1969 on a 12 acre (0.05 km^2) site on Cape Cod, US. Todd (1971), founder and co-director of the Institute, characterized what he calls biotechnology as being most effective at lowest levels of society, available to the poorest sections of society, advantageous to the development of decentralized communities, should use few resources, and be primarily based on ecological and social considerations rather than those of economic efficiency alone.

The New Alchemists, unlike many of the more utopian drop-outs, have been attempting to find economically efficient types of small-scale technology and have developed fish-farming using fish-ponds, and an 'ark' or glorified greenhouse incorporating a solar-heated and wind-powered food-growing complex. The balance of the ecosystem supporting the culture of fish and numerous fresh vegetables is closely monitored and regulated with the aid of a micro-computer. Far from a Luddite rejection of modern technology there is every attempt to harness it where possible to promote environmental and social objectives (Greene 1978; Todd 1976).

The promotion of alternative technology has now become more closely tied to the debate over nuclear power and the possibility of an energy gap. The view that alternative energy sources might come to our rescue is clearly a pragmatic response to concern about future energy supplies. It is a view that was pointed out as long ago as 1899 when Perry (in Flowers 1979: 53) in a book called *Steam Engine* wrote:

> For the last 20 years I have warned of the time when our stores of energy will be exhausted. By spending a few millions, nine-tenths of the energy in coal could be realised instead of one-tenth. When our store of coal is exhausted, the greater part of our civilisation will disappear. Then all places of high tide will become new centres of civilisation. Men will try to utilise stores of energy now thought to be insignificant – direct radiation from the sun, internal energy of the earth, wind power... There may be a new source of energy in a form unknown to engineers... If coal becomes more expensive, Lord Kelvin's idea of a reversed heat engine (i.e. a heat pump) will find favour.

Friends of the Earth and a number of conservation groups have paraded the virtues of alternative technology and conservation measures as an alternative to the development of nuclear power. Lovins (1977) recast the hard and soft technology distinction made by Clarke and others into a popular attack on conventional energy policy. Those concerned about the risks of nuclear power and the threat it poses to civil rights, nuclear proliferation, and centralization of decision-making have seen alternative energy strategies as a means both of preventing the horrific possibility of doomsday and of promoting democratic rather than corporatist politics. There has consequently been a curious alliance of social democrats, liberals, anarchists, Marxists and others in the promotion of alternative technologies.

Elliot (1978: 103), a long-standing supporter of alternative technology, commented: 'five years ago it [alternative energy options] was thought the preserve of well-meaning eccentrics and romantics. Now, however, the alternative 'renewable' resources... are beginning to attract the attention of energy planners and industrialists.' At much the same time a few of the older practitioners of alternative technology had begun to grasp some of the naïve assumptions about low environmental impact of windmills and solar collectors. Harper (1976) pointed out that many supporters of such technologies were unaware that their overall demand upon environmental resources, and the unnecessary work involved in their use, might be more than that of conventional systems. This was the argument taken up elsewhere by Inhaber in order to support the development of nuclear power (Ch. 2).

In 1973–4 when the oil crisis following the Middle East war brought home the dependence of the West upon oil, there was little corporate/industrial interest in alternative supplies of energy. Since then there has been a growing interest reflected by an increase in both public and private investment. By 1979 the UK solar heating market had crept up to a turnover of over £1 m. In 1978 government funds for wave power research were increased by £2.9 m. At the same time a £6 m. (over four years) solar energy programme was established. While this is still small compared with the £100 m. a year spent on nuclear power research it is likely to increase rapidly (Burge 1978). In the US the support for alternative energy supplies has been more impressive. In 1974 the government's solar budget of $14.8 m. was about 110 times less than its nuclear budget of $1.66 b. By 1978 the gap had closed considerably with solar power receiving $385 m. and nuclear power $1.36 b. (Bove 1979).

The extent of public and private interest in alternative technology and the way it has been manifested has caused some concern among the more radical supporters. Elliott (1979) expressed alarm at State sympathy towards the possibility of siting gigantic capital intensive offshore units to generate electricity from the waves for the national grid – a far cry from the self-sufficient ecological stability, and operational simplicity of the alternative technology movement. McKillop (1978: 44) another long-time supporter of alternative energy sources, concluded 'the vision of vast concrete booms, slung together like chains of upturned supertankers, and with hundreds of thousands of grid pylons and cables slashing from them across the Western Isles, is appealing to the 'mega-builder-technologist' that lurks so close below the surface of our energy officials and planners' (Fig. 3.2).

It is becoming clearer that an alternative technology future is possible without the social change implied by Clarke and other early protagonists. While Elliott, Lovins, Chapman (1976), and Leach (1979) all believe the possible energy gap in the future could be breached in the main by alternative energy sources other than nuclear power, they differ in their assumptions about the social change necessary. The low-energy future predicted by Leach as a viable scenario assumes that this is possible merely by a combination of conservation measures, product substitution, and technical

Prevention rather than cure: an alternative strategy

Fig. 3.2 The alternative mistopia. 'Come here this instant, or the Wave Power Generators will get you!' *Source: New Ecologist* (Mar/Apr. 1978: 43).

fixes. Conservation of energy plays a crucial role in his scenerio. He believed that a 50 per cent energy saving for space and water heating could be achieved simply by improving upon the present poor levels of thermal insulation. The iron and steel industry is capable of making a 30 per cent energy saving by improvements in blast-furnace operation, the gradual replacement of open-hearth furnaces by basic oxygen steelmaking, and heat-recovery techniques. Other improvements in fuel economy such as cogeneration (the simultaneous generation of useful heat and electricity) could have a substantial impact on conservation and at the same time help to prevent pollution.

The scope for conservation of resources in agriculture is great. There was excessive use in the 1950s and 1960s of cheap energy supplies, especially oil. Rising fuel costs since the energy crisis of 1973 have drawn attention to this issue. At the same time there has been a considerable waste of energy from power stations. Siting of greenhouses near power stations, and the use of low-grade heat could help conserve resources in the horticultural sector. Straw could be processed instead of being burnt in the fields, and organic fertilizers could be used as substitutes for chemical fertilizers (Fernie 1980). A shift in diet away from meat towards vegetarianism would also produce greater self-sufficiency and at the same time reduce energy demand (Mellanby 1975).

Conclusions

Policies involving alternative technologies, recycling and principles of 'pollution prevention pays' typically combine economic objectives with environmental objectives and are adoptable in a market economy. However, combination of the profit-maximizing requirement with pollution prevention and resource conservation can only be taken so far because of economic, administrative and political constraints. In a Marxist analysis the means of production, including technology, and its organization, are closely tied to the relations of production such as the interests of capital, domination, and hierarchy in a capitalist economy (Dickson 1974).

The alternative technology movement has offered an interesting set of moral imperatives and technological fixes, but the development of non-polluting technology is adopted only insofar as it makes sense within the constraints of the current demands of the economy. The adoption of some aspects of alternative technology, such as the social objectives outlined by Clarke, are antithetical to the interests of monopoly capitalism. Why, for example should multinationals and large companies which dominate economic activity wish to see the development of self-sufficiency and decentralization involving a fragmentation of capital and loss of power? The Chinese example, by way of contrast, illustrates how in a non-market economy alternative technologies and conservation can be integrated with economic and social objectives.

CHAPTER 4

International problems of pollution control

The growth in international trade, the development of chemical and other products which have pollution consequences transcending national frontiers, and the sheer quantity of pollutants being released into the environment have given rise to considerable international concern and action, especially during the last two decades. Holdgate (1979) has made a distinction between four categories of pollution which need to be dealt with at international level. There is inevitably some overlap between the categories, but they provide a useful framework for analysis. They are: 1. sources of pollution which originate in one country, cross national frontiers as a result of air and water currents and so affect adjacent territories – examples of 'transfrontier pollution' include pollution of the Rhine by Switzerland and Germany which affects the river entering into Belgium and Holland, and the sulphur dioxide emissions from Britain which are carried by the prevailing winds and are precipitated as acid rain over Scandinavia; 2. pollution arising from mobile sources such as aircraft, ships, motor vehicles, and products which have rapid dispersal characteristics such as pesticides, detergents, and pharmaceuticals; 3. the combined action of pollution from several nations may affect the equilibrium of common resources like the atmosphere, oceans and the global environment in general; 4. problems arising from different national approaches to pollution which may affect trade and other economic relationships between countries.

Transfrontier pollution

Transfrontier pollution leads to a conflict of interests between countries responsible and those that bear the consequences. National policy may aggravate the international problem: tighter restrictions on waste disposal on land may lead to a greater discharge into lakes and the sea; and regulations dealing with ground-level concentrations of sulphur dioxide may lead to taller chimneys and a wider dispersal of pollutants across national frontiers. For example, the British Clean Air Acts have enabled local authorities to force manufacturers to prevent local pollution by erecting taller chimneys. While reducing the urban sulphur dioxide problem in Britain this domestic policy has undoubtedly contributed, along with pollution from other European countries, to the acid rain problem in Scandinavia (Barnes 1979).

Transfrontier pollution may result from continuous emissions or accidental spillages, and may be the result of action from one or more countries. This gives rise to questions of liability similar to that resulting from pollution within national frontiers: either the polluter compensates the polluted or the polluted bribes the polluter to reduce his pollution. Resolution of a conflict of interests involves the establishments of property rights, accurate estimates of environmental damage and pollution control costs, and the development of institutions to reconcile international differences.

Principles governing the settlement of transfrontier pollution conflicts have been defined by international conferences, courts, organizations and commissions. However, no international agency exists which is empowered to give force to international environmental principles which are not incorporated into binding agreements. International legal principles do not as yet define the point at which a nation's interest in economic development must be modified to deal with transfrontier pollution. International environmental law is applied only when nations have consented to be bound by decisions of a neutral tribunal or commission.

The most notable case defining responsibilities of the polluter was the Trail Smelter Arbitration ruling in 1931. The conflict arose when a factory in British Columbia was found to be responsible for damaging vegetation and environment in the US. An International Joint Commission inquiry was established, under the Boundary Water Treaty 1909, which awarded damages against the Canadian firm. Compensation for damage was, however, restricted to established damages and did not include losses to amenity, which could not be estimated with any degree of accuracy. The Commission also laid down the principle that '... no state has the right to use or permit the use of its territory in such a manner as to cause injury by fumes in or to the territory of another or the properties of persons, therein, when the case is of serious consequence and the injury is established by clear and convincing evidence' (Wharan 1975: 268).

Many international organizations have attempted to formulate principles concerning the responsibility of states for transfrontier pollution damage. Principles 21 and 22 of the 1972 United Nations Conference on the Human Environment in Stockholm are exemplary (Rosencranz 1980: 16).

Principle 21 States have in accordance with the Charter of the United Nations and the principles of international law, the sovereign right to exploit their own environmental policies and the responsibility to ensure that activities within their jurisdiction or control do not cause damage to the environment of other states or of areas beyond the limits of national jurisdiction.

Principle 22 States shall co-operate to develop further the international law regarding liability and compensation for the victims of pollution and other environmental damage caused by activities within the jurisdiction.

The OECD Environmental Committee, while generally in favour of the 'polluter pays principle', recognizes the difficulties posed to a country which has a neighbour with very stringent standards. In some circumstances it may be fairer for the polluted country to participate in financial measures to reduce the levels required. Contrary to the strict liability principle it is debatable whether the polluting country should pay for the 'residual damage' that still occurs after the transfrontier pollution has been reduced to a mutually acceptable level (Smets 1976). A case can be made for cost-sharing on the basis that the costs of pollution are influenced by both the polluters and the polluted. Cost-sharing is most likely to be appropriate when a poor country with low pollution standards pollutes a better-off neighbour with more stringent standards.

The main principle adopted by the OECD in a ministerial level recommendation on transfrontier pollution in 1974 was of non-discrimination and equal rights of access (OECD 1976a). A good example of the application of this principle is the 1974 Nordic Environmental Protection Agreement between Scandinavian countries. The essential features of this principle are first that polluters causing or likely to cause damage in other countries should be subject to as stringent controls as those creating pollution solely within their own frontiers. Secondly, victims of pollution should not receive less compensation if their own national standards are less stringent than those of the polluter. Thirdly, an important feature of equal treatment is that victims should have access to administrative bodies and to the courts in polluting countries to ensure legal protection of their interests. Finally, the OECD believes that the principle should be applied not only to legal and administrative remedies but also to other rights such as access to information.

In effect the OECD has recommended that the victim of pollution should be able to seek a remedy to transfrontier pollution as if there were no frontier. In practice this can be achieved in three ways: 1. the victim could seek a ruling from a court in its own country; 2. the victim could seek a ruling from the country from which the pollution originated; and 3. the victim could seek a ruling from an international joint commission.

The first solution may not be feasible because of difficulties in executing a court's ruling. When agreements do exist between countries, as is often the case among OECD countries, courts may still be reluctant to grant injunctions, and in practice this alternative is mainly used for monetary compensation.

The second solution is likely to be difficult because victims of pollution in another country are in a comparatively weak position. There is a difficulty in organizing political pressure on the polluters, and a lack of information and familiarity with procedures. Access to the courts and administrative authorities is often difficult or impossible. Success is most likely amongst countries which have a history of close co-operation such as the Scandinavian countries, and countries of the OECD and EEC, but is less common amongst countries with widely differing political and judicial systems, levels of economic development and/or environmental policies. There have in fact been few instances when victims have sought a legal remedy in the country where the pollution originates. One of the main problems is the multiplicity of sources of pollution which makes it difficult to prove a claim and assign liability (Rosencranz 1980).

Lack of direct influence on the polluter often leaves the victim with little alternative but to press its own government to seek redress. Under these circumstances neighbouring countries may sign treaties and conventions or create bilateral and multilateral commissions to expedite matters. This third solution has been limited by the lack of suitable international commissions. However, a few do allow direct access to those affected by transfrontier pollution. The Finnish-Swedish Commission on Frontier Rivers is a case in point (Smets 1976).

International agreements

In 1979 the thirty-four member states of the Economic Commission of Europe signed a 'Convention on Transboundary Air Pollution'. The agreement included co-operation in monitoring and research, as well as a pledge to make efforts to reduce and prevent air pollution. The Indus Treaty in 1960 between India and Pakistan declared that both countries would try not to pollute each others' waters, and would not discharge untreated sewage or industrial waste in such a way that it would affect the use of water. Treaties and conventions, however, often only declare agreed intentions, and commissions often only involve advisory bodies and do not possess the executive powers to settle local transfrontier pollution conflicts. This conforms to good democratic principles where elected representatives have the power to make decisions since they are accountable to their electorate.

It is perhaps not surprising that many commissions and bilateral or multilateral treaties have had only limited success. In Europe the Rhine Commission was established in 1950 but this has made only slow progress towards agreements between nations on standards. The powers of the Rhine Commission, and similar bodies, have largely been confined to undertaking research and making recommendations. In 1978 delegates representing seventeen of the eighteen Mediterranean nations met in Monaco but failed to reach agreement on how to combat land-based pollution, which accounts for 80 per cent of pollution in the Mediterranean. Apart from a declaration of intention to do something about the pollution problem the only agreements achieved in 1978 were first to adopt a dumping protocol which would prohibit dumping certain poisonous chemicals ('black list') and require special permits for dumping less dangerous chemicals ('grey list'); and secondly to co-ordinate action in emergency situations involving accidents such as massive oil spills.

According to Müller (1979), bilateral or subregional agreements are most likely to succeed. He illustrated his point by referring to the successful international co-operation in recent years involving action by the US and Canada over pollution of the Green Lakes. In 1909 the two countries signed the Boundaries Water Treaty, which established a vague agreement not to pollute the Great Lakes in such a way as to cause injury to health or property. However, by 1972 pollution was so serious that another bilateral agreement, the Great Lakes Water Quality Agreement, was reached in order to improve the water quality. This Agreement has marked a significant development in bilateral co-operation. The International Joint Commission, established under the Boundaries Water Treaty, was required to monitor progress, and make recommendations. Between April 1972 and March 1973, known as the International Field Year for the Great Lakes, a $35 m. programme of research was undertaken in an attempt to quantify the pollution problems. The 1972 Agreement also required both countries to implement the following programmes (Müller 1979: 14):

1. completion of waste treatment facilities of all municipalities discharging into the Great Lakes;
2. installation of phosphorus-removal facilities of municipal sewage-treatment plants;
3. establishment of waste-treatment requirements for all industrial plants, including elimination of discharge of mercury, toxic persistent organic contaminants, radioactive materials, and thermal discharge;
4. measures for the control of pollution from agricultural, forestry and other land-use activities;
5. measures for abatement and control of pollution from shipping activities, dredging activities and onshore and offshore activities;
6. maintenance of joint emergency plans to deal with large oil spills and other hazardous polluting substances.

Not all of these objectives have been achieved and progress has been uneven. In 1978 a new Agreement was signed to continue the work started in 1972.

Perhaps the most comprehensive convention in the 1970s was the Helsinki Convention which was signed in March 1974 by all seven Baltic States. The Convention entered into force in May 1980, but during the previous six years an interim commission helped to expedite matters. The convention covers a comprehensive range of discharges from land-based sources and from ships, the deliberate dumping of wastes, and spillages. The Baltic States have agreed to take effective steps to reduce land-based pollution, and to counteract the introduction in the Baltic, by whatever means, of hazardous substances. Article 9 of the Convention prohibits the dumping of wastes except in emergencies (Hägerhäll 1980).

Mobile sources and mobile products

Manufacture of goods has become increasingly dominated by multinational companies. Improved communications and the growth of world trade have increased the transfer of goods. As a result there has been an increase in risks of transport which has international implications. Many synthetic chemicals which fail to break down rapidly are also highly mobile and toxic, causing problems that are difficult to tackle at a national level only.

Transportation of oil

Without effective international agreement, transport activities will take place under regulations drawn up by countries with less stringent controls. Thus 'flag of convenience' countries such as Liberia, Panama and Cyprus account for twice, six times and eight times respectively the world average loss rate of ships. Seventy-five per cent of world vessels lost are registered with these three countries and Greece (Nagel 1978).

Each year about two billion tonnes of oil are transported by sea. The size and total tonnage of tankers has increased enormously, especially following the closure of the Suez Canal in 1967. The world production of oil also doubled between 1965 and 1977. By 1974 there were nearly 400 oil tankers of 200,000 deadweight tonnes or more. The first major incident was the *Torrey Canyon* disaster in 1967 involving a cargo of 119,000 tonnes of Kuwait crude oil. This led to the majority of its oil (95,000 tonnes) escaping from the scene of the accident – the Seven Stones Rocks between Lands End (England) and the Isles of Scilly. As a result 25,000 seabirds were killed at the Rouzic in Brittany alone. Since then there have been several notable oiltanker accidents. A most serious recent casualty was the *Amoco Cadiz* in March 1978 when 230,000 tonnes of oil spilled into the English Channel 3 km from Brittany's north-west coast.

Some of the oil entering the marine environment is associated with marine accidents, collisions, and groundings, but a much greater amount comes from marine operational losses, offshore production, and land-based discharges (see Table 4.1). In July 1979 the worst oil spill to date occurred with a blow-out from a well being drilled in the Bay of Campeche by the Mexican National Oil Company. Estimates of the oil spill, which continued

International problems of pollution control

Table 4.1 Sources of oil in the marine environment

Sources	Tonnes
1. *Marine operational loss*	
Tankers	1,000,000
Bilge discharges	300,000
2. *Marine accidental discharges*	
from all sources	350,000
3. *Offshore*	
Production	150,000
Natural seepage	600,000
4. *Land based discharges*	
Refineries, petro-chem. plants	
and waste oils	1,300,000
Total	3,700,000

Estimate of the amount of oil entering the marine environment in a typical year (say 1976 from all sources).

Source: Wardley-Smith 1979: 11.

unabated for several months, ranged from 10,000 to 30,000 barrels per day.

Accidents and the problems posed by increasing transport of oil and other cargoes have led to a number of international conventions and agreements. Discharges from tankers are controlled by international agreement through the Inter-Governmental Maritime Consultative Organization (IMCO). In 1954 thirty nations signed an Agreement not to discharge ballast tanks, containing seawater and oil, in specified areas. The limitations of this agreement were that the areas covered were too restricted and several countries with large tanker fleets did not sign it. In 1962 the Agreement was extended to cover a wider area and more countries signed it. In 1969 there were further amendments which imposed more stringent requirements such as the oil content of discharge and the distance travelled by the tanker during discharge. In February 1972 twelve north-west European states made an agreement in Oslo dealing with the dumping of hazardous wastes into the North Atlantic. Another convention in the same year in London established a similar agreement amongst a greater number of countries. In Britain the Dumping at Sea Act 1974 gave effect to these two conventions. In 1973–4 the north-west European states went a step further when a convention in Paris agreed to control the discharge of noxious wastes into the sea from the land via rivers, pipes and coastal outfalls.

Discharge at sea of oil-contaminated ballast water is not necessary as tankers can be loaded with oil using the load-on-top technique whereby ballast water is drained off at the oil-storage terminal while oil is pumped in on top of the ballast water. The process continues until the first signs of oil in the ballast water appear. Nevertheless some tanker captains prefer to discharge ballast water at sea and run the risk of detection. As a consequence there is often a severe destruction of wildlife. Sage (1979) has pointed out how this illegal practice began to devastate Shetland's wildlife soon after the opening of a huge oil-storage and tanker-loading complex on the island.

There is a very great difficulty in enforcing international discharge regulations. France uses a monitoring programme involving aircraft to spot oil discharges, and evidence for legal action is collected by infra-red photographs of the ship's wake. Use of satellites to detect infringements of the law is another possibility. In the UK the 1977 Annual Report of the Advisory Committee on Oil Pollution of the Sea recorded that there had been 432 oil-pollution incidents, of which the source was identified in 124 cases, and there were fifty-one prosecutions leading to fines of £60,000. The main difficulty, it was concluded, was the cost of monitoring and the relatively low levels of fines (O'Riordan 1979). The problem of enforcement costs is discussed in greater detail in Chapter 6.

Following the *Torrey Canyon* incident IMCO states concluded an agreement in 1969 which gave states the right to take measures to stop pollution of the seas when it threatens their coast. Various safety procedures have been worked out by the Maritime Safety Committee of IMCO. These include provision of sea lanes, additional navigational aids, shore guidance systems, speed restrictions, periodic testing of navigational equipment, training of officers and crew, and the construction of tankers (Central Unit on Environmental Pollution 1976a). Unfortunately there is still a failure by some nations to ratify and enforce existing conventions, which has allowed substandard ships and improperly trained crews to continue transporting vast quantities of oil. For example, the 1974 Safety of Life at Sea Protocol when in full operation would reduce the number of substandard ships and improve crew training, but at the time of the *Amoco Cadiz* incident only thirteen countries had ratified the Protocol (Nagel 1978).

Although there has been a somewhat limited response to international conventions, tanker owners are faced with liability for oil pollution. An International Convention of Civil Liability for Oil Pollution Damage in 1969 agreed that there should be strict liability for oil spillages from tankers in all convention countries. Moreover, two voluntary schemes have also been adopted to deal with the costs of oil pollution – the Tanker Owners' Voluntary Agreement Concerning Liability for Oil Pollution (TOVALOP) and the Contract Regarding Interim Supplement to Tanker Liability for Oil Pollution (CRISTAL). The former agreement covers liability in case of negligence (about 70 to 80 per cent of marine accidents result from human error. The remainder of causes include technical problems such as failure of steering equipment or navigational equipment.) The latter involves participating oil companies providing a Compensation Fund for cases not involving negligence (McLoughlin 1976).

Mobile products: the case of PCBs

Control of polychlorinated biphenyls (PCBs) illustrates the international problem of pollution control from products. PCBs, like certain pesticides such as DDT, are extremely persistent, mobile and toxic. They are to be found long distances away from the source of manufacture and use. They have been found in brown seals off the coast of Scotland, shrimps in Florida, cod in the Baltic Sea and mussels in the Netherlands. They are transmitted widely through the oceans and the atmosphere. Their effect as pollutants is most obvious on wildlife, where they cause brittle birds' eggshells. In many respects control of PCBs is more difficult than DDT which has restricted use as an insecticide. PCBs are widely used in epoxy paints, protective coatings for wood, metal and concrete, and carbonless reproducing paper, but probably their largest use is as coolants and insulators in high-voltage transformers. When PCBs in rubbish are incinerated they vaporize and form a gaseous pollutant which is precipitated in rain. PCB pollution also results from manufacturing and waste disposal. Hence if PCBs are not regulated at an international level they will continue to move across national frontiers and affect those nations that have banned their use. Thus the US Toxic Substances Control Act of 1976 which banned the use of PCBs in January 1979 and banned the production of PCBs in July 1974 can only be regarded as a limited although important step in the right direction. As Jackson (1971: 243) commented,

> ... if PCBs are still disposed of carelessly on a substantial scale in one part of the world, wild life throughout the world remains in danger. If DDT is banned in the Northern Hemisphere but not in the Southern Hemisphere the whole world will continue to be affected. In these and presumably in many other potential situations of a similar character, international solutions are the only effective remedy.

Pressure to regulate mobile products has been limited and has depended chiefly upon international efforts to monitor pollution at a global level. In the 1960s and 1970s there was a growth in international scientific appraisal of pollution problems by such organizations as the United Nations' Environment Programme, the World Health Organization and the World Meteorological Organization, as well as by professional groups such as the Scientific Committee on Problems of the Environment. The EEC has also made some progress towards co-ordinated control through its Dangerous Substances Directive. This requires member states to comply with quality objectives, for a limited number of substances, based on toxicity, persistence and accumulation of substances in the environment.

Regional and global pollution problems

Most pollution problems are localized. Microgeographical variations in bronchitis, for example, may reflect different levels of smoke and sulphur dioxide concentrations (Ch. 1). However, sometimes the effect of several sources of pollution leads to regional problems. Confined common resources such as the Great Lakes in North America and the Mediterranean Sea are examples that have already been mentioned. The Baltic and the Irish Seas are other reasonably land-locked seas where pollutants discharged by several countries can accumulate. In the Irish Sea, 12,000 seabird deaths have been attributed to the accumulation of polychlorinated biphenyls (Holdgate 1979). Discharges from various heavy metals threaten wildlife over large areas of ocean. This threat occurs not just because of their mobility and persistence, but also because of the accumulation and their toxicity in low concentrations. Radioactive contamination from nuclear explosions could also have devasting regional effects.

Regional and global climate may be affected by pollutants, especially carbon dioxide from the use of fossil fuels. It has been suggested that increased levels of carbon dioxide will lead to a 'greenhouse' effect (allowing incoming radiation from the Sun but reducing radiation back into space) with a reduction in heat loss from the Earth and a rise in world temperatures. This could lead to a gradual melting of Greenland and Antarctic ice with a consequent rise in sea levels so altering coastlines, fishing and climate. Furthermore, chlorofluorocarbons that find their way into the stratosphere may reduce heat loss. It has also been suggested that the amount of radiation reaching Earth will fall, due to increased aircraft vapour trails and particulate levels in the air. On the whole, however, the evidence for changes in global climate resulting from pollution is inconclusive.

Although knowledge of regional and global effects is at an early stage of investigation, Holdgate (1979) points out that there has been a considerable increase in international scientific effort to monitor possible effects. The United Nations' Environmental Programme's Global Environmental Assessment (Earthwatch) in particular has the function of monitoring, information exchange, research, evaluation and review.

Trade and economic relationships

Differences in national standards of pollution control have considerable implications for international trade. They can lead to distortions in costs of production, to artificial trade barriers, and to the transfer of hazardous manufacture to so-called 'pollution havens'.

About four million chemical substances have been identified, of which about 30,000 are commercially produced, but there are many more waste products and intermediates. Consequently something in the vicinity of one million chemicals are involved in the process of manufacture, use and disposal of products. Each year 500 to 1,000 newly synthesized chemicals reach the production stage in the US alone. The long-term effects of low concentrations of these products are often unknown for twenty or thirty years. Possible effects include genetic mutations, cancer and birth defects (UNEP Annual Report 1978). Up to 25 per cent of the research and development budget of a modern chemical corporation is spent on researching into possible environmental risks associated with a major new product. In the case of a new pesticide this can amount to as much as $20–25 m. (O'Riordan 1979).

Given the costs and extent of research necessary to assess environmental risks, international agreements on methods of experimental techniques and exchange of information about new chemicals are clearly both desirable and necessary to avoid disruption of trade. Without such agreements countries might think it prudent to ban the entry of products which do not conform to their own standards. Banning of products on grounds of safety or pollution may of course be little more than a deliberate trade barrier in disguise. As Johnson (1978: 270) commented on the need for harmonization within the EEC: 'Barriers to trade will clearly be created if widely different conditions of use relating to new substances are imposed in different community countries.'

There would be a wasteful duplication of research effort if information on experiments to test the safety of products was not made internationally available. The difficulty is that a firm will probably have spent a considerable sum of money in developing a new product, and information concerning its properties may involve disclosure of valuable commercial data. With a rise in concern about product safety there has been a growth in information systems. Port (1978) has described the UK system (DESNET), a European system based in Italy (ECDIN) and a global system based in Geneva (IRPTC). Much more information is likely to be stored at each level when compulsory notification to governments about new chemicals becomes more widely practised.

The OECD has been particularly involved in harmonizing the regulation of chemicals in member countries. This has become increasingly important since some countries have introduced tougher measures to regulate use of dangerous substances. The consequences of the US Toxic Substances Control Act 1976, as Draggan (1978: 262) argued, will include a considerable impact on international trade in chemicals. It has, he pointed out, been 'characterized nationally and internationally by some manufacturers as a definite barrier to trade'.

Strict pollution control standards may help to protect a country from competition abroad. Detailed regulations on motor vehicle performance, especially concerning exhaust emission, have helped to prevent imports of foreign cars into Japan. Another feature of strict standards is the export of polluting industry to countries with less stringent controls. Multinationals' choice of industrial location will take into consideration environmental costs as well as many other aspects of manufacture such as labour availability, quality and costs; public services, transport costs, proximity to markets; the nature of the firms' own logistic network, character of product; social and political structure of the country, the nature of investment controls; tax policies; and the availability of raw materials and intermediary inputs within the productive system. For most industries the costs of pollution controls are usually of minor importance in comparison with other considerations. However, for certain industrial processes involving significant risks the consideration of pollution control costs may be a crucial factor in industrial location.

Some Japanese industries, especially when involving petroleum products, have been established in nearby countries because of lack of suitable sites and costs of pollution control in Japan (Walter 1975). Several oil refineries, petrochemical works, and pharmaceutical plants that serve the US economy are located in Puerto Rico because of pollution problems (*Science for the People* 1973). Nedlog Technology Group Inc. of Colorado has offered to pay $25 m. a year to ship toxic wastes to Sierra Leone because of the high costs of 'safe' disposal in the US (*New Scientist* 7 Feb. 1980).

It is often argued, especially amongst several EEC countries, with the notable exception of the UK, that fixed emission standards would prevent distortions in trade. However, a strong case can be made for emission standards which vary according to characteristics of the local environment. The costs of similar quantities of emissions will vary from one location to another, and so it is reasonable to apply different emission standards. Pollution control costs are, as stated, just one of many variable costs in production. It seems reasonable that industry should be encouraged where costs, including environmental costs, are least. This is one of the reputed advantages of international trade. Nevertheless, differences in standards for certain products and emissions can lead to serious abuse and distortions of trade. The task ahead continues to be one of harmonizing international policy before more independent national regulatory actions become widespread.

Conclusions

This chapter has striven to demonstrate the need for international co-operation and agreements in order to control the damage caused by pollution. Regulations to control pollution within one nation often have repercussions for other nations. Very stringent standards in industrially developed countries alone give rise to a very real danger of less developed countries becoming the production centres for highly polluting and risky industries. Attempts to control local impact from mobile pollutants within one nation may lead to wider dispersal and harmful consequences for a neighbouring country. Independent national pollution control policies may also lead to artificial trade barriers. International co-operation has become more and more necessary to protect common resources such as rivers and seas from pollution hazards involved in the transportation of resources and products. The last decade has witnessed gradual progress towards an understanding of these issues, and a number of international agreements have resulted.

CHAPTER 5

Institutional and administrative arrangements for pollution control

Introduction

This chapter will survey some of the main features of administration in the UK, the EEC and elsewhere. Typically there are numerous authorities responsible for pollution control, each with its own geographical boundaries, often determined by historical accident as much as by environmental and demographic characteristics of the area. Pollution control administration varies significantly from country to country. An international comparison will be made of some aspects of decision-making which can be characterized in terms of polar opposites such as open-closed, centralized-decentralized and corporatist/consensus-adversary systems.

Lutz (1976) has distinguished between different types of environmental institution. At central government level there are co-ordinating councils, environment ministries, agencies having independent status, and ministries which deal with environmental matters among other responsibilities. In Australia there is an Environment Council which co-ordinates environmental legislation, assesses research requirements and makes recommendations. In the Federal Republic of Germany there is a cabinet committee for the environment. Several countries including the UK, Australia and Canada have created a ministry predominantly responsible for environmental affairs. In the US, on the other hand, the Environmental Protection Agency is more independent than a ministry in the sense that the agency head cannot be removed for political reasons. Japan also has an independent Environment Agency. In Sweden there is no single environment ministry or agency, and the Ministry of Agriculture is the main government authority responsible for environmental policy-making.

Central government agencies are supported by a variety of advisory committees which help to provide information and organization of data to aid decision-making. For example, the Central Council for Environmental Pollution Control advises the Japanese National Environment Agency. In

61

the US, the Council on Environmental Quality advises the President on environmental affairs and produces a detailed annual report on the state of the environment. National, regional and local institutions manage environmental affairs according to the type of problem and locality, as well as the extent of decentralization within the political system of the nation concerned. Local authorities are usually the most appropriate level of administration for the more common and straightforward pollution problems where familiarity with local conditions is important. Regional bodies are more appropriate when pollution has a regional significance.

Administrative arrangements in the UK

The legislative and administrative procedures for pollution control in the UK have been reviewed at length by the Central Unit on Environmental Pollution (1976b). At the central government level pollution control is shared by several ministries. The Department of the Environment in conjunction, as appropriate, with the Scottish and Welsh Offices is concerned with air and fresh-water pollution, noise pollution other than from aircraft, oil and chemicals on beaches, and disposal of solid and radioactive wastes. The Department of Trade is responsible for the control of marine pollution by oil, and other forms of pollution including noise and emissions from civil aviation activities. The Department of Industry is responsible for several research establishments that investigate pollution problems, including the Warren Springs Laboratory which does considerable research on air pollution, oil pollution at sea, and industrial waste treatment and recovery. The Ministry of Agriculture, Fisheries and Food is responsible for controlling the use of pesticides, and in order to protect fisheries is also responsible for controlling marine pollution and dumping at sea. The Department of Energy in conjunction with the UK Atomic Energy Authority and the Central Electricity Generating Board is responsible for controlling nuclear wastes. The Health and Safety Commission, which is responsible for health and safety at work, includes the Factory Inspectorate and the Alkali and Clean Air Inspectorate. The Department of Health and Social Security and the Scottish Home and Health Departments advise other departments on the health hazards posed by pollutants.

One of the implications of horizontal fragmentation of responsibilities is the need for some overall co-ordination of pollution control between departments. This is the responsibility of the Secretary of State for the Environment. He is assisted by directorates forming the Environmental Protection and Water Engineering Command. The Central Unit on Environmental Pollution is involved in making appraisals of pollution problems, and in providing management capability for an overall monitoring and assessment system. The standing Royal Commission on Environmental Pollution set up in 1970 also studies problems referred to it by Ministers and advises on all matters relating to pollution. There are several other government bodies which provide advisory services. These include the Advisory Committee on Pesticides and other Toxic Chemicals, the Noise Advisory Council and the Clean Air Council. The latter, which was set up under the Clean Air Act 1956, reviews the progress in abating air pollution in England and Wales. It also gives advice. Its recommendations concerning the publication of information were embodied in the Control of Pollution Act 1974.

At the regional level water-pollution control is the responsibility of ten water authorities in England and Wales. They were established under the Water Act 1973 and are both autonomous and financially self-supporting. The National Water Council acts as an advisory body and provides a liaison between the water authorities and the Department of the Environment.

Local government has responsibility for several pollution problems including the control of air pollution from domestic, commercial and industrial sources which are not the responsibility of the Alkali Inspectorate. Local authorities are also responsible for controlling noise, for waste disposal, and in coastal areas for dealing with oil pollution. Harbour authorities are responsible for dealing with pollution in harbours.

Most of the wide array of governmental organizations responsible for pollution control in England and Wales are illustrated in Fig. 5.1. The practical application of air and water pollution control policy will be discussed in greater detail in Chapters 6 and 7. This sketch of administrative responsibilities in the UK illustrates the fragmentation of responsibilities despite the existence of an environmental ministry. It also illustrates the horizontal and vertical divisions of responsibility typical of environmental administration. The extent of these divisions varies mainly according to the degree of centralization and importance attached to environmental affairs from one country to another.

International comparison of pollution control policy-making and administration

Corporatist/consensus-adversary systems

Several authors have compared different styles of environmental policy-making and administration. On the one hand, according to O'Riordan (1976 and 1979), there are those countries where decision-making takes place following confidential discussion between selected groups that are considered trustworthy. This 'consensus approach', he claims, is typically found in the UK, some Commonwealth countries and some countries within the EEC. On the other hand, there are those countries where debate is open, public suspicion prevails, and many negotiations are conducted through lawyers. This 'adversary approach' is typical of the US, Japan and

Institutional and administrative arrangements for pollution control

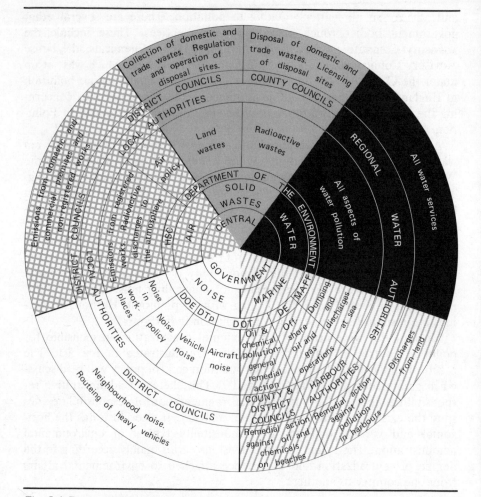

Fig. 5.1 Responsibilities for pollution control in England and Wales. *Source*: Central Directorate on Environmental Pollution.

West Germany. Decision-making in Britain is therefore based more upon consultative procedures than is the case in the US, compromise and conciliation being established British practice.

Decision-making does not often fall into quite such black and white categories and the situation is obviously fluid. O'Riordan (1979), for example, referred to the costs of the American adversary approach typified in environmental impact statement and risk assessment procedures. As a consequence, he suggested, there has been an interest in finding more acceptable corporatist strategies. Likewise, although the UK relies heavily upon administrative discretion, quasi-judicial procedures have been adopted under pressure from environmental groups. At the 1977 Windscale public inquiry into a planning application from British Nuclear Fuels to build a nuclear reprocessing plant a judge, Justice Parker, presided over proceedings. Both British Nuclear Fuels and the environmental lobby in opposition made considerable use of barristers in presenting their case.

Open/closed systems

Kelley et al. (1975) offer a similar open/closed classification of types, but their categories depend more upon possibilities for public participation and scope for access to the political system by environmental groups than upon the adversary/consensus distinction. They did, however, draw a distinction between consensus and conflict-orientated systems. In their terminology conflict was interpreted as occurring in the pluralistic competition of interest groups, and consensus was used to describe systems where such conflict was largely absent. They characterized environmental policy-making in the Soviet Union and Japan as depending upon relatively closed and consensus-orientated decision-making systems, and that of the US as a relatively open and conflict-oriented process.

Because of the closed characteristics of policy-making in the Soviet Union and Japan, environmental groups have according to Kelly et al. (1975: 132) 'found it difficult to organize effectively, either because of the formal restrictions on the organization of independent groups as in the Soviet Union or because of informal pressure not to upset the delicate (and economically profitable) political balance as in Japan'. As a consequence, environmental groups tend to form around single-issue affairs such as the pollution of Lake Baikal in Russia or the mercury pollution tragedy of Minamata in Japan. In Japan environmental groups have achieved most success by using the courts. By 1975 there had been four major pollution-related injury cases involving itai-itai disease from cadmium pollution in water, minamata disease in two localities from mercury pollution in water, and asthma from air pollution involving sulphur dioxides (Nomura 1975–6). Hence environmental administration in Japan can be described as both closed and adversary in character.

In the closed system power tends to be concentrated in the hands of the elite. In Israel, for example, environmental policy is usually initiated within the Knesset and the role of pressure groups is small. In contrast to the multitude of environmental groups in the US and the UK there are in Israel only four environmental associations. Only one of these can be considered to be a grassroots organization, and even this is not really concerned with environmental politics. The other three groups were established by the Knesset, by officers within the Ministry of Health and by senior scientists (Yishai 1979).

One of the features of an open system of decision-making is that representatives of environmental groups or non-governmental experts on pollution are often appointed to advisory boards or as consultants to government agencies that deal with pollution problems. Public participation often takes place during public hearings, which play a significant role in US decision-making, or public inquiries which are a key feature of land use planning in the UK. Environmental groups can also exert an influence through comments on government reports, submissions to agencies on pollution problems, briefing of sympathetic politicians, and liaison with administrators and the mass media (Kimber and Richardson 1974).

The US system of policy-making is regarded as open because rival interests can organize and lobby with few restrictions. Policy outcomes tend to reflect the power and influence of contending groups. The main federal agency is the Environmental Protection Agency (EPA) which was established in 1970 and is responsible for research, monitoring, standard-setting and enforcement functions. There are also various other agencies with environmental responsibilities, such as the National Forest Service, Soil Conservation Service, Army Corps of Engineers and Bureau of Reclamation (Kelley et al. 1976). Many of the states also have agencies principally concerned with environmental affairs such as the New Jersey Department of Environmental Protection. There are also some fifty interstate agencies such as the Coastal States Organization, whose principal functions are to develop a common approach and to act as a clearing house for information.

The institutional structure in the US allows many points of access by interest groups. Kelley et al. (1976) suggested that there are 250 national and regional organizations dedicated to environmental protection. The National Center for Voluntary Action has estimated that some 5.5 million people belong to or contribute to nineteen so-called primary environmental organizations. There may be as many as twenty million active environmentalists attached to the 40,000 environmental/conservation voluntary organizations (Fanning 1975). Business and industrial lobbies are organized primarily along functional or industry-based lines, and include the oil lobby, highway lobby, and the agricultural lobby. These lobbies have considerable advantages as they often have control over information, and have a working knowledge of formal and informal systems of regulation. Environmental groups, on the other hand, depend upon information dissemination in order to exert an influence at public meetings, discussions and public hearings, as well as in lobbying, in litigation and in protests and demonstrations. There is therefore often an imbalance of power when environmental groups such as the Friends of the Earth, Environmental Action and the Sierra Club take on industrial interests such as the National Association of Manufacturers. Moreover, the apparent openness of the policy-making process does not necessarily mean that industry has lost its power to structure pollution-control policy and administration in its favour (Ch. 7).

Richardson (1979) has described the Swedish policy system as principally devolved, rationalistic, consensual and open. Central ministries are small and a good deal of decision-making is devolved. Planning is usually based on extensive prior research and regular use of 'study commissions'. Policy decisions are typically reached after considerable agreement of all parties concerned. During the passage of the Environment Policy Act 1969, which covers water and air pollution and noise from stationary sources, there were conflicting views as to whether the Swedish pollution control system should be consensus/corporatist or more adversary in character. As a compromise, two agencies were set up to administer pollution control. Polluters must either apply for a permit from the National Franchise Board, which is court-like and staffed by lawyers, or gain an exemption from the National Environment Protection Board. The two boards are altogether independent of each other. The Franchise Board must, according to the law, balance economic and other benefits of an activity against established or potential damage to the environment. When the damage is considered unacceptable but, nevertheless, the activity is considered of great importance the Franchise Board may leave the case to government decision. Both the applicant and the Environment Protection Board have the right to appeal to the government against a decision by the Franchise Board (Bouveng 1980).

Documentation of the types and quantities of pollutants involved, together with the permits or exemptions issued, are usually published in full and are freely available for public inspection and comment. Enforcement is the responsibility of the National Environment Protection Board, but in practice the Board plays mainly an advisory and co-ordinating role. The twenty-four county administrators are the principal controlling authorities.

Lundqvist (1979) has compared the US and Swedish approaches to air pollution control. The Swedish approach has been based on standards that are practicably obtainable given current technology and economic considerations. The US legislation, on the other hand, was a more hasty response to environmental concern in the late 1960s and was based primarily on public health considerations with little regard to economic and technical feasibility. Standards were aiming at future targets which were beyond current possibilities. Nevertheless, as Lundqvist points out, in the US average emissions per vehicle kilometre decreased by 18 per cent between 1970 and 1976. In Sweden there was only an 8 per cent decrease over the same period, which was more than eliminated by a 25 per cent increase in vehicles. The pressure of a more adversary system with the constant threat of litigation may have been the crucial factor. However, there is still a problem of non-compliance in the US. In the case of water pollution, the EPA estimated that 14 per cent of major industrial discharges did not achieve the 'best practicable technology' statutory deadline of 1 July 1977. But a study by the Water Pollution Control Federation estimated that only 33 per cent of municipal wastes complied with secondary treatment requirements by the same date (Council on Environmental Quality 1977). The problems of pollution control enforcement will be dealt with more fully in the next chapter.

Centralization versus decentralization

One of the main differences in pollution control policy-making and administration is that between centralized countries like France, Finland and Sweden, and federal countries like the US and West Germany. In the US the states issue permits but the Federal Environmental Protection Agency has the right to veto them. In the Federal Republic of Germany, central government is even less involved. It only lays down general directives, whereas the Länder deal with detailed implementation and enforcement.

According to Lutz (1976), federal governments with weak central government such as Australia, Canada and the Federal Republic of Germany have faced constitutional difficulties when confronted with environmental problems of national concern. In Germany conflict was resolved by constitutional amendment to allow federal control of ambient air quality and waste disposal. In Canada the provinces have retained independent powers over environmental affairs, but the national government has gained powers to take action in emergency situations or in situations involving interprovincial or international problems.

The Soviet system of policy-making has been characterized as one that is highly centralized and geared principally to the single-minded drive for industrialization (Gustafson 1978). Responsibility for environmental affairs is fragmented among numerous agencies. Most national policy-making of importance takes place within the Politburo. High-level negotiations have generally paid little attention to state agencies and semi-official environmental groups. There is a lack of political mechanisms allowing concerned citizens to use the media or mobilize public opinion or resort to the courts, administrative hearings or parliaments to bring pressure upon the polluters as in Western democracies (Kelley *et al.* 1976). Industrial interests have tended to block the influence of environmental groups. Even in the late 1960s there was only a weak system of planning and enforcement of air and water quality standards. In 1970 only 16 per cent of municipal wastes throughout Russia were being treated. In 1972–3 only one quarter of industrial wastes and process water discharged into the Volga was being treated (Gustafson 1978).

The low priority of environmental protection and the pollution consequences have been comprehensively described by Goldman (1972), and Kelley (1976). It was in response to a deteriorating environment, and in line with international concern during the late 1960s and early 1970s that Russia promoted pollution control to the front line of national policy-making. In 1974 environmental protection was made an official part of the national economic plan, and a special division for the environment, with subdivisions responsible for air and water quality, was created in the State Planning Committee. In financial terms the support for investment in water quality jumped in 1973 from 300 to 1,500 m. rubles per annum. Then in 1976 Brezhnev announced a five-year, 11 b. ruble programme of capital expenditure for protecting the environment, most of which was to be devoted to preventing water pollution.

The Ministry of Reclamation, known as the Minvodkhoz, is now the agency responsible for water use and quality. Despite its annual budget of 7 b. rubles a year, the Minvodkhoz is not as powerful as other ministries with which it must contend. These are the ministries of the chemical industry, oil refining industry and ferrous metallurgy industry. The river basin inspectorates, which are accountable to the Minvodkhoz, are responsible for enforcement, but are not allowed to interfere with industrial objectives involving high-priority output targets nor with local employment. On the industrial side there are few incentives to do work on pollution control. According to Gustafson (1978: 470) 'Much of the power of the basin inspectorates is of the wrong kind. Their sanctions apply primarily to local enterprise directors, who lack the means to comply. They do not reach the real sources of the problem: the construction contractor, the R & D and design institutes, the supply organizations, and the higher levels of the ministries.'

Whether or not pollution control decision-making is centralized or decentralized may be influenced by constitutional considerations, but also by the nature of the pollution problem. In the UK responsibility for air pollution control is split between the centralized Alkali Inspectorate and local authorities. The history of administration of water resources has been a move away from ad hoc and laissez-faire development of bodies responsible for water supply, sewage disposal and enforcement of regulations to the development of regional unitary bodies responsible for the whole water cycle. Edwin Chadwick, writing in the mid-nineteenth century, expressed the view that the supply of water, sewerage, drainage, cleansing and paving should be under one and the same local management and control. Progress towards this objective has, however, been slow, and it is only in recent decades that the interconnectedness of water management and river pollution has been fully recognized.

The problems of providing an unspoiled water supply for domestic, industrial and agricultural consumption are closely related to a number of interconnecting factors such as water availability, water demand, pollution, and administrative control over the whole water cycle. With a growing demand for water resources for domestic, industrial and agricultural consumption more and more water must come from rivers which have already received discharges from industry and sewage works.

In 1963 in England and Wales there were twenty-nine river authorities responsible for setting and enforcing consent conditions. There were 500 water undertakers, half the number there had been in 1945, and there were numerous sewage works administered by about 1,400 local authorities. Pressure for reorganization grew towards the end of the 1960s and recommendations for reorganization were outlined in the Working Party on Sewage Disposal Report (1970) called *Taken for Granted*, and a Central Advisory Water Committee Report in 1971. The main problem with the fragmentation of responsibilities was that local authority sewage works were polluting rivers which were used by other private and local authority water undertakers. There were few political votes in sewerage and costs of pollution were often felt by other interests. Improving sewage works was not therefore of great priority. The Working Party on Sewage Disposal recognized this problem and recommended that there should be an integration of sewerage and water functions.

Standard-setting and enforcement by river authorities solely responsible for controlling river pollution and developing water resources should, on the face of it, have led to effective control. However, river authorities were dominated by local authority representatives. As a consequence river authorities rarely prosecuted local authorities, no matter how overloaded or outdated were their sewage works. The Report of a River Pollution Survey of England and Wales 1970 (1971) indicated that 37 per cent of all discharges of sewage effluent or 60 per cent of the total sewage effluent discharged exceeded standards required by the river authorities. About half of the industrial discharges also failed to meet prescribed standards. The report estimated that the cost at constant 1970 prices of bringing the country's rivers up to the standards then expected by the river authorities by 1980 would amount to about £610 m.

These problems led to pressure for reform and the Water Act 1973 reorganized the water industry and brought into being on 1 April 1974 ten new water authorities. These are responsible for water supply, waste disposal from sewage plants, standard-setting and enforcement. The water authorities cover as far as possible whole river-basin systems. As a consequence there is now much less inclination to allow costly pollution from sewage works. The creation and development of river-basin authorities in France, the Netherlands and other countries have had a similar purpose of integrating responsibilities for different demands upon the water cycle (Barde et al. 1979).

The water authorities in England and Wales have faced a difficult problem in taking on the responsibilities for both sewage disposal and enforcement of consent conditions originally imposed by the river authorities. Overnight the water authorities became the greatest abusers of standards and at the same time became responsible for enforcement. Government restrictions on capital expenditure compounded the problem. Between 1973 and 1975 there was a 53 per cent drop in municipal water and effluent treatment plant ordered for the home market, and in 1976 there was a 30 per cent cut in water authorities' capital expenditure. As a consequence water authorities continue to abuse their own standards based in the main upon principles set out in the eighth report of the Royal Commission on Sewage Disposal in 1912. This situation has put the water authorities in a weak position to prosecute industry for non-compliance with consent conditions.

In 1976 the National Water Council recommended a solution to these difficulties. It suggested a review of consent conditions according to the demands upon the river system. In areas where there were many discharges and there was increasing recycling of water, standards should be more severe. The counterpart to this was a relaxation of standards where use of the river system required less stringent standards of water quality. The water authorities accepted the recommendations of the National Water Council and have since been revising their consent conditions (Sandbach 1977a and b; Bugler 1978; Bates 1979; Parker and Penning-Rowsell 1980).

Another problem with the regional autonomy of water authorities is the lack of central direction. With a growth in demand for water there are likely to be long-term shortages of water in regions with a low rainfall such as East Anglia. In March 1976 the Labour government strongly condemned the lack of central guidance. Regional shortages of water could be solved in part by the transfer of water through canals, but this is only practicable if the quality of rivers is satisfactory and if there is an acceptable inter-regional plan of action. The drought in the summer of 1976 drew attention to regional problems and the dangers of pollution to rivers with much lower flows of water. However, economic circumstances prevented a further reorganization of the water industry (Elkington 1977; Grove-White 1976).

The EEC's approach to pollution control

In the EEC the Council of Ministers is the sole source of legislative authority. It is supported by the Commission, which consists of thirteen members appointed by the governments of member states. There are two members from the UK, France, Germany, Italy and one member from each of the other states. The Commission has a large staff (equivalent to a national civil service) of some 7,500 people of all Community nationalities. Some of the administrative powers of decision have been delegated to the Commission both by the Treaty of Rome and by the Council of Ministers. The Commission is also responsible for drawing up proposals to be considered by member states in the Council, and for implementing policy decisions. The role of the European Parliament, now composed of elected representatives from member states, is limited to offering opinion and advice. It has no legislative or executive authority.

The Commission of the European Communities works on proposals for the following types of action: *Regulations* which are applied directly by the EEC; *Directives* which are binding, but are applied by member states; *Decisions* which are binding on the parties concerned; and *Resolutions* and *Recommendations* which have no force. The Council of Ministers cannot act unless there is unanimous agreement amongst its members. While this is a safeguard to national sovereignty, it does mean that considerable debate and compromise must take place before agreement is reached. This involves discussion between representatives of all nine member countries

Institutional and administrative arrangements for pollution control

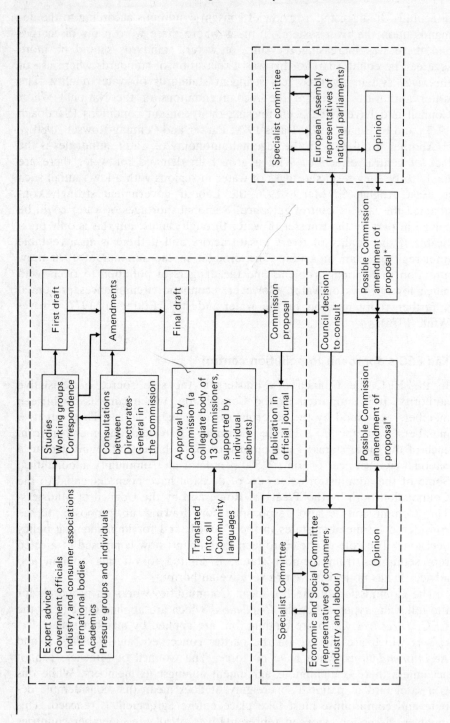

Centralization versus decentralization

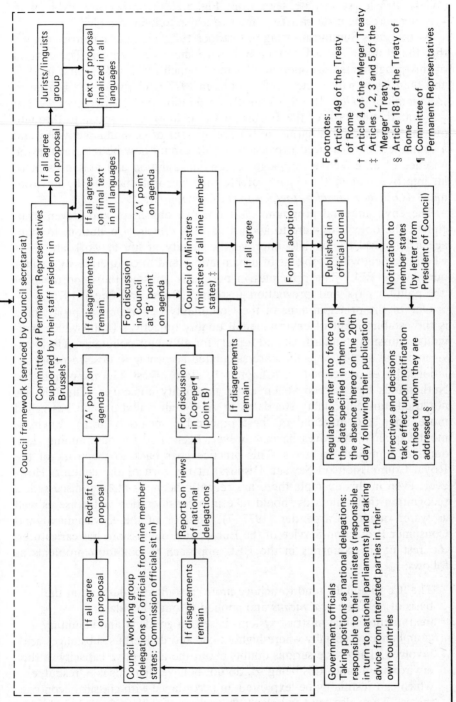

Fig. 5.2 Processes involved in the making of a directive. *Source:* Renshaw (1978).

(Wilde 1978). As can be seen from Fig. 5.2, the processes involved in securing agreement on an EEC directive are extremely complex.

At the Paris Summit meeting in October 1972, member countries agreed that there should be a Community Environment Programme. The content of this programme was approved by the Council of Ministers, first in 1973 and then again with some modifications in 1977 (Bulletin of the European Communities 1976). In the Council's declaration of 22 November 1973 it was stated: 'the task of the European Community is to promote throughout the Community a harmonious development of economic activities and a continued and balanced expansion, which cannot now be imagined in the absence of an effective campaign to combat pollution and nuisances or of an improvement in the quality of life and the protection of the environment' (*Official Journal of the European Communities* 1973: 1–2).

The environmental programme set out five objectives and eleven principles. The objectives were to abolish harmful effects of pollution; to manage a balanced ecology; to improve the quality of life at work and elsewhere; to deal with problems of urbanization; and to co-operate with states outside the EEC. The principles in particular stressed the importance of 'the polluter pays' and 'prevention at source' (Booth and Green 1976).

One of the principal aims of the Community's environment programme is the establishment of environmental quality objectives (upon which individual emission standards should be set) for all major pollutants. These environmental quality objectives are calculated according to dose/response relationships as described in Chapter 1. Italy, West Germany and the Netherlands have already been using air quality objectives. The UK has not used ambient quality standards at all for air pollution and has relied more upon the principle of 'best practicable means' (Ch. 6). Environmental quality objectives have, however, been used for determining standards of emissions to rivers. This practice goes back to the work of the Royal Commission on Sewage Disposal at the turn of the century. However, even in this example there has recently been, as already discussed, a recognition that standards should take into consideration water use as well as water quality. Carpentier (1977: 7), director of the Environment and Consumer Protection Service of the European Commission, has caricatured the rest of the countries in the EEC approach to pollution problems as follows:

> The 'Continentals' tend to believe more in standards defined on the basis of *best technical means* and applied through mandatory instruments. They mistrust systems based on goodwill and voluntary compliance, especially where highly toxic and dangerous substances are involved. They have serious doubts about the absorptive capacity of the environment, and, this being so, do not believe that this is a 'resource' which can justifiably be 'exploited' in accordance with classical economic theory. They also tend to believe that where the scientific and medical

evidence is inconclusive, we should err on the side of caution.

On the whole Britain has pressed the EEC to adopt a pollution control strategy based on guidelines, because of her special situation of having considerable opportunities for dispersing pollution, particularly in estuaries and coastal areas. Other EEC countries, especially Germany, have claimed that this would lead to lower standards in Britain which would give her firms an unfair trading advantage. Such differences in approach to pollution problems have led to some conflict and misunderstandings within the EEC.

Between 1973 and 1979, fifty-six proposals came before the Council. Forty-one of these have been adopted (Mandl 1979). Among draft directives which were opposed by the UK was one setting a standard for lead in the air. It was opposed on grounds of lack of scientific evidence. Another seeking to control the sulphur content of oil was opposed on the grounds that high sulphur content of oil was not always the cause of pollution problems (Wilde 1978). When in 1976 the UK Royal Commission on Environmental Pollution made its fifth report on methods of controlling domestic and industrial air pollution, there were proposals from some quarters for the EEC to adopt rigid ambient standards with penalties for non-compliance. The Royal Commission argued that ambient standards were generally impracticable because of the difficulty of relating them to specific sources. The best practicable means approach was favoured because it was more flexible and took into consideration local financial and environmental implications. Concern for health had always been considered in the operation of best practicable means, but it was admitted by the Royal Commission that insufficient attention had been paid to air quality. It was in favour of establishing guidelines which would not have legal force but would be considered in setting limits on emissions. These guidelines would indicate an upper level above which action to reduce concentrations should be taken and a lower level below which such action would not be justified (Flowers 1977). By the end of 1976 it was clear that the EEC would not have secured agreement over enforceable ambient standards for air pollutants, and as a compromise air quality objectives in line with the Royal Commission recommendations are being adopted.

Although the UK has been in favour of quality objectives and against setting fixed emission and ambient standards for the majority of pollutants, it has agreed to a more rigid system of controls on dangerous substances. The Directive of 4 May 1976 on dangerous substances discharged into the aquatic environment cleverly steers a course which is acceptable to the different types of administrative practice within the EEC. The first directive on dangerous substances had only proposed a system of controlling 'black list' substances by emission standards. However, the British favoured an approach to water pollution problems based on setting environmental quality objectives, and the draft directive was rejected. The Commission was then given the task of working out a compromise between the emission standards favoured by the majority and the quality objectives approach favoured by the UK (Commission of the European Communities 1977b).

The final directive on dangerous substances placed an obligation on member states to control discharges of 'black list' substances by the application of uniform emission standards laid down by the Council, unless it could be shown that quality objectives, also determined by the Council, are met and maintained. These quality standards are based on the toxicity, persistence and accumulation of the substances in living organisms and in sediment. The Commission has chosen the following substances from the 'black list' for priority action: mercury, cadmium, aldrin, dieldrin and endrin (Commission of the European Communities 1977b). For 'grey list' substances, it will be up to member states to authorize emission standards based on quality objectives (Otter 1979).

The main role of the EEC in pollution control has been one of harmonizing policies of member countries when agreement is possible and worthwhile. According to Thairs (1977: 28) the greatest scope for action is provided by 'the establishment of common environmental quality objectives; the alignment of research programmes; rationalization of the methodology required to assess the progress being made, or likely to be made, to protect and improve the environment; and the systematic and co-ordinated exchange of information'. These and other proposals for action have been set out in the environment programmes of 1973 and 1977–81. Additional strategies worthy of note are those aimed at the harmonization of specifications for products, such as motor cars, which cause pollution, studies of pollution problems in individual industries such as the paper and pulp industry, and measures to harmonize rules for implementing international conventions. The 1977–81 programme particularly emphasized the importance of setting up the machinery for preventing pollution and the production of wastes.

Conclusions

In this chapter a comparison has been made of different aspects of pollution-control administration. Some features of the decision-making process can be related to ideological and constitutional characteristics of the State. Other features have developed as a pragmatic response to characteristics of the pollution problem. Hence regional water authorities and river basins have been created within a great variety of political systems. The EEC's approach to pollution control has been described. The principles and objectives of the EEC Environment Programme have led to some conflict among member states, particularly over the debate about the merits of ambient standards and quality objectives. The main task of the EEC is, as was demonstrated in the previous chapter on international problems, that of harmonizing different countries' approaches for mutual benefit.

CHAPTER 6

Economic considerations in pollution control and the regulatory approach in practice

Introduction

In Chapter 1 it was suggested that an efficient pollution control strategy should attempt to balance the costs and benefits of control. Techniques for doing this were shown to be more or less comprehensive. After collecting information on the impact of pollution and costs of control it is possible to decide how much reduction in pollution is desirable. Some empirical evidence suggests that the marginal costs of pollution increase as pollution increases. That is to say the costs of each additional unit of pollution tend to be more than those of the previous one. There is also some evidence that the marginal costs of pollution control increase (Kneese and Schultze 1975). A grossly simplified diagram (Fig. 6.1) illustrates changes in the marginal costs of pollution as pollution increases from 0 to Z, and the marginal costs of control as pollution is reduced from Z to 0. Let us assume, again for the sake of simplification, that we are only concerned with the emissions of one firm. According to principles of economic efficiency, there is an optimum amount of pollution control. In this abstract case pollution should be reduced to a point P. Any less pollution control would result in marginal pollution costs exceeding marginal costs of control. Any further pollution control would result in marginal costs of control exceeding marginal pollution costs. Neither situation would be desirable if the only goal of pollution control is to maximize benefits minus costs.

The optimum level of pollution control can be achieved, it is claimed, either by setting a standard of 0P pollution or by levying a tax 0T per unit of pollution. The argument behind the tax solution is that a firm would wish to reduce its tax burden (equivalent to area ABZP) by paying less on pollution control (equivalent to area APZ) (Kneese et al. 1970; Beckerman 1973 and 1975; Marquand 1976). In principle the same result could be achieved by bribing polluters to reduce their pollution. In the example illustrated in Fig. 6.1, if a bribe of 0P per unit of pollution controlled were

Economic considerations in pollution control

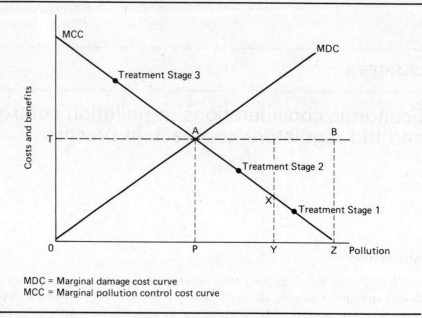

MDC = Marginal damage cost curve
MCC = Marginal pollution control cost curve

Fig. 6.1 The optimum level of pollution control without threshold.

applied it would be in the interests of a firm to reduce its pollution to the optimum level 0P, but no further.

The bribes approach is disliked for several reasons. First, it goes against the 'polluter pays' principle (OECD Environment Directorate 1974). Secondly, the pollution charge makes the firm pay the full social cost of pollution and hence the price of a product will include both production and pollution costs. Thirdly, bribes may act as an incentive for firms to increase pollution in order to qualify for bribes (Beckerman 1975). However, bribes might well be justified in certain situations. Pollution costs are not created by the polluter alone. When people come to live near a source of pollution it can be claimed that they have caused the extra inconvenience and extra costs of pollution, rather than the polluter. In such situations, if a higher standard of pollution control is demanded, then it may be fairer for the community as a whole to bear the costs and bribe the polluter to reduce pollution. Similarly, if an industry is attracted to an area on the assumption that pollution standards are not a major constraint to production, and at a later stage the community wants higher standards, it might be more appropriate to bribe the firm.

In practice subsidies and grants are often used in conjuction with a standards/enforcement strategy. In Britain, under the Clean Air Acts of 1956 and 1968, householders are entitled to a 70 per cent grant towards the cost of installing smokeless appliances when they are obliged to do this under a smoke control order. In Japan there are generous depreciation allowances and tax concessions for expenditure on industrial pollution control technology. In the US subsidies for the construction of waste-water treatment plants have been available since the 1956 Water Pollution Control Act Amendments. One of the limitations of these tax subsidies and grants is that they have tended to encourage narrowly defined methods of control rather than a choice of strategy from a wider range of options (Council on Environmental Quality 1976).

Much debate has focused on what are the advantages and disadvantages of pollution taxes and standards. Some of the debate has centred on theory whilst some has been concerned with the operation of standards and taxes in practice, and has emphasized the role of pressure groups and the organization of the economy. Before reviewing these arguments it is worth noting some fundamental limitations of the optimum pollution control model which are relevant to the theoretical debate.

It was pointed out in the first chapter that there are numerous ways of controlling pollution. Emissions control might not be the most efficient approach. It may, as has been suggested, be easier to reduce the costs of pollution by taking protective measures. The pollution control model described only balances the costs of pollution against the costs of reducing levels of pollution. By way of contrast, the Common Law of nuisance has in many situations recognized the importance of balancing costs and benefits between the polluter and activities effected by pollution. Some economists believe that the efficiency criteria which guide Common Law are more flexible than those adopted by Statutory Law (Posner 1972; Coase 1960). However, transaction and litigation costs in the control of pollution and the various problems encountered when defending individual rights to a clean environment in a complex society have made Common Law of limited value in pollution control (Sandbach 1979).

A second problem with the optimum pollution control model is in drawing the graphs. The steady increase in marginal pollution costs ignores the possibility of thresholds to damage (Chapter 1). If thresholds exist these may be the most important factors in determining optimum pollution control levels (Fig. 6.2). Representation of marginal pollution control costs as a continuum can also be criticized. There may be a variety of possible methods of treatment but these may not fall into a continuous range. There is for example, a big jump in costs as one moves from primary to secondary and tertiary methods of sewage treatment (Martindale 1979). The optimum level of pollution control should consider the costs of moving from one treatment stage to another. In Fig. 6.1 the optimum pollution control level 0P may in fact be impossible to achieve if there is no available treatment method between treatment stages 2 and 3.

A third problem with the model is that it fails to take account of enforcement costs. If the model is adopted without consideration of these costs the so-called optimum level of pollution control may be at a level which is either impossible or difficult to enforce. The enforcement costs are

Economic considerations in pollution control

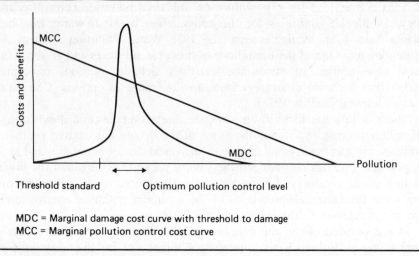

Fig. 6.2 The optimum level of pollution control with threshold.

likely to vary with different levels of control and ought to be considered when formulating pollution control policy.

A fourth problem with the optimum pollution control model is its indication that the optimum level exists at the point P where the benefits of pollution control minus the costs of pollution control are maximized. However, as one approaches the so-called optimum point the benefit/cost ratio approaches the figure of 1. In other words, at point P the returns on investment in pollution control are just equalled by the benefits. Most expenditure, whether it be in the private sector or public sector, is expected to give much higher rates of return. The actual optimum amount of pollution control in terms of opportunity costs (the return on investment spent elsewhere) may be YZ in Fig. 6.1 where the benefit/cost ratio for an extra reduction of a unit of pollution is 0T/XY. Bearing in mind the difficulties associated with an optimum pollution control model, we can now review the professed advantages and disadvantages of a pollution tax strategy, and of a standards/enforcement strategy.

The advantages of pollution taxes

1. The innovative incentive

The use of a pollution tax has a distinct advantage when the problem is the lack of suitable means of control. The firm is encouraged to find a cheap way of reducing pollution control in order to reduce the tax burden. In a simplified abstract example, in Fig. 6.3, $0P_1$ represents the optimum level of pollution control. However, if the firm finds new methods of reducing

The advantages of pollution taxes

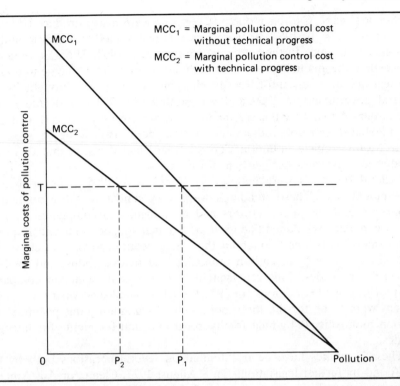

Fig. 6.3 The innovative incentive of a pollution tax.

the marginal costs of pollution control, it is able to reduce the level of pollution to $0P_2$ and at the same time reduce the pollution tax costs to the firm. Critics of the standards/enforcement approach have often referred to the lack of incentive for innovations in pollution control technology. There is an incentive to meet a given standard at least cost but there is no incentive to search for new technology which would reduce pollution further as is the case with pollution tax. While the innovative argument favouring the operation of pollution taxes has not as yet been supported by the use of taxes in practice, it can be supported in a more limited sense by reference to delays in innovation when controls are exercised through use of standards. Bugler (1975) believed that the lack of innovative incentive is the reason why a major firm like London Brick, with a £25 m. annual turnover, took little initiative in looking for improvements in pollution control, but waited for new technology to be innovated elsewhere.

The practical problem of encouraging innovations through the use of standards is illustrated by long delays in achieving effective controls on motor vehicle exhaust emissions in the US. The first effective controls were introduced in California where legislation succeeded in forcing car manufacturers to provide exhaust-control devices. Hydrocarbon controls were introduced on 1961-model cars and were made more restrictive for

1966-model cars. Carbon monoxide controls were also introduced for the 1966-model cars. In 1964 car manufacturers had argued that they would be unable to introduce further effective controls until 1967. There is some evidence that the manufacturers had colluded to prevent implementation of the legislation, as an anti-trust suit charging collusion was brought by the federal government in 1969 and was settled by a consent decree in the same year. Apparently the car manufacturers were only induced to implement exhaust emission reductions on 1966-model cars while in 1964 the State Motor Vehicle Pollution Control Board had declared that there existed at least two practical systems for doing so.

Federal standards were first set by Congress in 1965 to be applied to 1968-model cars. These standards were very similar to those applied in California and car manufacturers had no difficulty in complying. In 1970 the Clean Air Act Amendments took a much tougher line and required that exhaust emissions throughout the US be reduced to 90 per cent of the 1970 levels by 1975 for carbon monoxide and hydrocarbons, and by 1976 for nitrogen oxides. This represented a 97 per cent reduction compared with uncontrolled conditions. In 1971 the EPA established what these regulations were to be for the three pollutants: 3.4 g/mile (grams per mile) for carbon monoxide, 0.4 g/mile for hydrocarbons and 0.4 g/mile for nitrogen oxides.

These deadlines have been extended on several occasions both by the EPA and by further legislation. On 8 August 1977, Clean Air Act Amendments postponed the enforcement of the final hydrocarbon standard to 1980-model cars and that of the carbon monoxide and nitrogen oxides standard to 1980-model cars. The nitrogen oxides standard has also been relaxed from 0.4 g/mile to 1.0 g/mile. The history of delays illustrates the difficulties of legislating for future standards based upon uncertain and unknown technology. Industry was given no incentive to search for improvements as would have been the case had there been a tax reduction incentive (White 1976a and b).

2. Allocative advantages

A pollution tax has an advantage if there are several discharges in the same locality, and if the firms concerned have different marginal costs of pollution control. In this situation, when a charge is applied, the dischargers (with no financial constraints) would tend to reduce pollution to a level where the marginal cost of abatement equals the charge. A tax solution would ensure that those who find it cheapest to reduce pollution do most of the pollution control, while those who find it more expensive will do less pollution control. This is illustrated in Fig. 6.4 where we have two firms A and B producing the same levels of pollution Z with different marginal costs of pollution control. If a tax level 0T is applied firm A, with more favourable abatement costs, will control more pollution than firm B

The advantages of pollution taxes

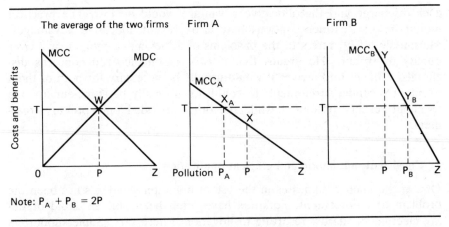

Fig. 6.4 The allocative advantage of a pollution tax.

with less favourable abatement costs. Firm A will reduce pollution to P_A and firm B will reduce pollution to P_B (where $P_A + P_B = 2P$). If both firms had to comply with a standard 0P then the extra costs of control to firm B would have been more than the reduction on costs of control to firm A.

An interesting example of the allocative advantages of a pollution tax would arise if there were a tax on motor vehicle exhaust emissions. Instead of requiring all cars to maintain a particular standard there may be allocative advantages in ensuring that heavy-car users have controls, but that occasional car users are exempted. For occasional users the cost of special exhaust systems is likely to exceed the social costs of pollution. Kneese and Schultze (1975) have suggested a tax solution to the problem. If cars were tested periodically for the volume of exhaust emitted, they could then be given a smog rating to determine a level of tax which would be collected on the purchase of petrol. The occasional user would be more prepared to pay a high tax on petrol rather than spend more on controlling exhaust fumes. Although the administrative costs of issuing smog ratings and collecting taxes are a drawback these are likely to fall dramatically with the rapid growth in computerization and 'electric' money which we are now experiencing.

If a standards approach were to obtain the same allocative advantage as a tax then it would be necessary to have precise knowledge of each firm's abatement costs. There is serious doubt as to whether this could be determined with any accuracy except by the firm itself. Moreover the administrative costs of setting many different standards based upon allocative considerations and the possible sense of injustice, militate against such an approach. The allocative advantages of taxes are less when discharges take place in different conditions, and where different standards are necessary. In such circumstances it would be necessary to vary the levels of tax for each discharge and the allocative advantage would be lost. The practical limitations of treatment options may also prevent allocative advantages. Martindale (1979) refers to the problems of achieving category 1 or 2 river quality in Britain. He argues that if two reasonable-sized concerns discharged effluent into a river it would usually be necessary for both of them to adopt secondary treatment; it would not normally be sufficient for one to adopt tertiary treatment while the other provided only primary treatment.

3. Consistent, automatic and efficient application

One of the main weaknesses in the use of pollution standards has been the problem of enforcement. Polluters have often been able to avoid or delay meeting standards, resorting to litigation if necessary. Some supporters of pollution taxes claim that the application of a tax would immediately result in a response, and that it would be less prone to political fashions (Beckerman 1975). The financial burden of a tax is considered to be more effective than the pressure to control pollution through the threat of fines, imprisonment, or even the threat to the public image following an adverse court action. While this is not the view of industry (Martindale 1976), it is supported by the difficulties faced in introducing and enforcing legislation in most countries. On the other hand, the record of introducing pollution taxes fails to support Beckerman's views. Wenner (1978) and Majone (1976a and b) have reviewed the scant empirical evidence relating to pollution taxes. The operation of pollution taxes in France, Netherlands and Vermont State in the US has been subject to protracted discussions and had at time of review failed to impose an efficient solution to water pollution control. However, the levels of pollution tax have been increasing over time and the intention is that they should reach an incentive level in the long run (Barde *et al.* 1979). The view of Majone and Wenner is that different enforcement mechanisms, whether they be fines or taxes, are subject to similar political pressures, which limit their effectiveness. Consequently different pollution taxes and standards are unlikely to have substantially different results if the same distribution of political power exists in the communities concerned.

An illustration of the way pollution charges, let alone taxes, have been affected by industrial pressure is discussed by Freeman (1977). In Britain the Public Health Act 1973 enabled industry to discharge its waste directly into the public sewers, and at the same time gave local authorities the right to impose a charge for this service. The industrial lobby, through what was known as the Group of Prescriptive Right Discharges, ensured that legislation exempted from charge all those industries that had been discharging effluent for a minimum of a year prior to the Act. Moreover, local authorities were not in a position to charge the true costs of treatment because of the inter-war unemployment problem and the competition between local authorities in attracting industry. Even in 1977 only 60 per cent of the treatment costs in the Severn Trent Water Authority Region were met by industrial discharges.

In a review of water pollution charging schemes in the EEC, McIntosh (1978) found that active pollution tax systems were applied only in France and the Netherlands. In neither case were the tax levels introduced at a sufficient level to act as an incentive to reduce pollution efficiently. In both countries it was decided to start with a low level of taxes and then steadily raise them in order to let firms get used to the idea and adjust. The taxes helped to provide a revenue which was used to reduce the burden of environmental improvement. Italy and Germany were also in the process of introducing similar types of pollution taxes. Belgium and Luxembourg were considering the introduction of a scheme, while Ireland, Denmark and the UK had made no plans for introducing pollution taxes. The nearest the UK has come to introducing pollution taxes is in enabling sections 52 and 53 of the Control of Pollution Act 1974 which give the Secretary of State the powers to permit water authorities in England and Wales and river purification authorities in Scotland to introduce taxes (Martindale 1976) but so far these powers have not been applied.

In another review of pollution taxes, Barde *et al.* (1979) summarize features of the tax system as applied to water pollution control. None of these systems rely exclusively upon charges. In France, the Netherlands, West Germany and Hungary, charges are backed up with a system of direct regulations. In Hungary fines or taxes are levied on firms and municipalities for any of thirty-one specified pollutants which exceed prescribed levels. In the Netherlands taxes are assessed on measured chemical oxygen demand and the extent of nitrogenous substances. In France taxes are assessed on measured suspended solids and oxidizable matter.

The disadvantages of pollution taxes

Martindale (1976) has claimed that there are four major disadvantages of a pollution tax system: complexity, uncertainty, insensitivity and expense. These disadvantages, like the advantages, are dependent upon the particular problem of pollution control, and sweeping generalizations must therefore be avoided.

1. Complexity

A pollution tax is most appropriate in the situation where the pollutant is easy to monitor and the tax is easy to apply. In the case of exhaust emissions a tax could be applied fairly easily at the source of manufacture. White (1976b) argued that this way of tackling air pollution would have been preferable to the standards control policy adopted in the US. It would, he claims, have led to better research and faster emissions control. On the other hand, a pollution tax is less likely to be appropriate for dealing with complex industrial processes. According to the UK Alkali and Clean Air Inspectorate, pollution taxes may be appropriate for simple and non-fluctuating pollution emissions, but in the case of industrial air pollution control, such a system would be unworkable because of the complicated emissions system. Any workable system, it argues, must 'embrace imponderables such as meteorological variations, indirect emissions such as leakages, breakdown, wind-whipping, good house-keeping, maintenance and all other items previously described which cannot be measured... Why replace the existing simple effective system with a complicated, unproven theory?' (Department of the Environment 1973: 15.)

Martindale (1976) argued that when it is necessary to have a great variety of standards operating in different areas and for different pollutants associated with effluent discharge then a standards approach is easier to administer. In the UK, the government resisted an EEC move towards uniform emission standards for discharges, believing that a more flexible approach was required which would take into consideration local environmental and economic circumstances. The more flexible and complex the requirements of pollution control, the less appropriate a tax system becomes. Martindale cited the OECD's Environment Directorate which, on review of the scope for pollution taxes, concluded that it would be too complicated and costly to have a charge for each and every pollutant.

2. Uncertainty

Several factors lead to uncertainty about the effect of administering pollution taxes. Storey and Elliott (1977) argued that there will still be difficulties in estimating the effects of a tax even when the marginal costs of pollution control are known. They suggested two principal reasons for uncertainty. First, improvements in pollution control require capital expenditure, but this is more likely to be limited by internal rationing within the firm than is revenue expenditure. The second reason is that a firm's decision whether or not to spend money on pollution control may be influenced by other than financial considerations. These reasons can be expanded on. Insights into the behaviour of a firm are particularly important. Given limits upon capital expenditure which prevent firms from investing in everything they would like to, then choice of investment alternatives will be influenced by the expected rates of return on investment. If a firm invests in pollution control it may not gain as much revenue as it would from investment elsewhere (Sandbach 1979). Different firms would react in different ways, and as Martindale (1979) argued, the amortization of capital expenditure over different periods would produce significant variation in cost as between one discharger and another.

For some economists uncertainty about the effect of a pollution tax is unimportant, for if the tax were applied incorrectly then adjustments, or 'fine-tuning' of pollution control, would be possible by altering the tax rate until the desired level of pollution control had been achieved (Beckerman 1975). However, changing tax rates in this manner gives rise to some serious problems. In particular it increases uncertainty for the firm. In this situation a firm may simply delay taking a decision to see what change in tax rate is made. If this is done to any great extent it would make fine-tuning of tax levels inherently unstable and unsatisfactory. If on the other hand the majority of firms respond to a wrong level of tax they may well have to pay more in the long run than the firm that delays taking action until the tax rate has settled down. The reason for this is to be found in the way pollution abatement costs vary according to the strategy of implementation. Pollution control technology is not always additive. A move to a higher level of pollution control may involve the scrapping of previous methods of control and their replacement with more effective measures, or it may involve a redesign of the manufacturing process. Storey and Elliott (1977) claimed that the wrong level of tax would tend to 'lock in' a firm to the wrong level of technology. Alternatively a firm might adopt a second-best strategy of choosing a system of pollution control which is flexible but less efficient. Burrows (1974) argued that more information on the marginal costs of pollution control is required to administer pollution taxes than pollution standards, in order to avoid the trial and error procedure and the 'lock-in' effect. Standard setting has the advantage over pollution taxes in reducing uncertainty.

3. Insensitivity

Chugh et al. (1978) believed that the flurry of pollution control legislation in the US at the beginning of the 1970s increased firms' business and financial risks. A system of taxes would, if set at an effective rate, impose an even greater burden upon firms as they would have to pay a pollution tax as well as pay for controlling pollution. Pollution taxes would allow less room for making exceptions and they would hit the small firm in particular as there are often substantial economies of scale involved in pollution control. Large quantities of pollution are often relatively cheaper to control and in some instances may give rise to useful resources which would not have been worth collecting in smaller plants. Large sewage works, for example, as in the case of the West Middlesex and Mogden works, produce methane which is used to provide electricity. Small zinc and lead smelters which emit low levels of sulphur dioxide are in a much weaker position to control this effluent than large smelters which can viably convert gases into sulphuric acid.

Alkins and Lowe (1977) claimed that large firms can raise finance more easily from external sources because investors' risks are lower, and they can also raise finance more easily from internal sources through cross-subsidization within the corporation. The large firm usually has better access to other resources such as technical information, which helps to make a more efficient response to demands for pollution control. Alkins and Lowe, on the basis of a survey of industries in the east Midlands, believed that uniformly enforced standards have tended to hurt small firms, especially at a time of recession and low profitability. A uniformly imposed tax system would have hurt the small firm to an even greater extent.

It might be argued that the most efficient firms can survive extra pollution control costs and that the less efficient firms do not deserve to survive – a policy of allowing lame ducks to go to the wall. However, the cost of closing a firm could be much more harmful than the costs of the pollution whose control is sought (Martindale 1976). The question also arises whether it is desirable to operate an inflexible pollution control strategy that favours the large firm in this way. Small firms are often more innovative and have greater entrepreneurial skills despite their less sound financial basis. The majority of the Royal Commission on Environmental Pollution (Third Report) in 1972 believed that an efficient system of pollution taxes would cause unwarranted harm to the small firm. A standards/enforcement strategy is able to balance the financial difficulties of the firm, the costs of pollution and the contribution of the firm to the local community. A more flexible policy of enforcement than would be the case for pollution taxes is possible for if tax exceptions were made the whole system would probably become unworkable. Certainly some of the claimed advantages of a pollution tax would be lost.

4. Expense

In many circumstances the monitoring and administrative costs of a tax system will exceed those of a standards/enforcement system. Common (1977), in recognition of these costs in setting taxes, has suggested that a tax on inputs rather than emissions would be easier to administer. He suggested that a tax on the sulphur content of oil would be an effective way of reducing sulphur dioxide emissions from fuel-burning. The problem with taxing inputs is that it only encourages substitution of inputs but not the conversion of harmful emissions into something useful or a change in combustion technology which reduces sulphur dioxide emissions. An emissions tax or standard encourages a wider choice of pollution control strategies. However, even a pollution tax on emissions would discourage techniques of pollution control which divert the pathway of a pollutant from sensitive targets or protect targets from pollution (Ch. 1).

Martindale (1976) argued that a pollution tax would be more expensive to a discharger than would a standards/enforcement system, and hence would be more inflationary. Nevertheless, as has already been stated, if a pollution tax is equivalent to the external costs of pollution then the price of a product is more likely to reflect the true cost of production and not just the internal costs to the firm, which would occur without a tax. Although in theory the costs of pollution taxes could be neutralized by a reduction of other taxes, doubts as to whether this would be done lead to strong industrial opposition to pollution taxes (Martindale 1976 and 1979; Buchanan and Tullock 1975).

Low levels of pollution tax, which are not calculated to induce efficient strategies of pollution control, in conjunction with standards of pollution control have some financial advantages. The pollution tax would go some way to ensure that the firm pays for the cost of pollution and that at the same time a tax revenue is provided which can be used to finance the pollution control administration or to compensate those adversely affected by pollution. A redistributive tax system can be used to subsidize those firms that are required to do most of the pollution control in order to meet a quality objective. The extra costs to those firms that treat more effluent are lessened by a contribution from firms that treat less. It is this kind of low-level tax system which has been most commonly adopted. In the Netherlands, State and regional water pollution taxes are collected. State taxes are used to further investment and running costs of private plants; regional taxes are used for financing regional plants and other public expenditure (McIntosh 1978). In Finland a small charge is levied on liquid effluents and this is used to pay administrative costs and to increase the stock of fish (Lutz II 1976).

The case for and against pollution taxes is a complex one, with advantages and disadvantages depending upon the type of pollutant and the discharge situation. A blanket policy with no flexibility in the application of pollution taxes would seem to be undesirable. The EEC Commission have been looking at pollution tax proposals, and according to one proposal there have been moves to require all EEC countries to introduce a tax scheme for all direct discharges to water systems including surface and underground water, and the sea. The taxes would be set at a sufficiently high level to act as an incentive to control pollution. Revenue would also be used to improve water quality (Martindale 1979). Such blanket recommendations fail to take into consideration the disadvantages and limitations of pollution taxes for many forms of pollution. The case against standards/ enforcement and for pollution taxes is not as black and white as several commentators (Beckerman 1975; OECD 1974; Majone 1976a and b) have made out. While much of the debate about pollution taxes has been theoretical in nature, with only few empirical cases to draw upon, there is a wealth of evidence to illustrate the adoption and enforcement of pollution standards. The rest of this chapter will look at some of the features of the direct regulations and enforcement approach associated with the use of pollution standards.

Features of the regulatory approach

In a review of water pollution control policies, Barde *et al.* (1979) identified four key features of the regulatory approach. First it is the oldest and most widespread approach. This is true unless one considers the regulatory control of pollution through the Common Law of nuisance, trespass, negligence and strict liability. It is certainly true that the majority of countries have come to rely upon statutory direct regulations. The second feature is the administrative, legalistic and judicial character of standard-setting and enforcement. A third feature is that permits or consents to discharge are granted on a case-by-case basis. Guidelines set at national and regional levels help to define what levels of pollution are permissible. The fourth feature is that permit systems rarely operate without some form of charge, bribe or tax relief system. An additional and important feature is the need to have an effective means of enforcement.

Enforcement problems

The costs of enforcement and the costs to the firm of avoidance measures have been inadequately studied. Nevertheless some of the major issues have been treated by economists including Becker (1968), Stigler (1970) and McKean (1980). While the enforcement authority will have to spend money on enforcement, those being regulated may well invest time and money on avoidance measures such as seeking loopholes in the law, delaying compliance with new standards, through litigation if necessary, and concealing offences. The significance of enforcement costs is shown in the considerable outlays involved. In 1977 the US Federal Government spent about $400 m. on the standard-setting and enforcement for pollution control. McKean (1980) has reviewed some of the enforcement difficulties which should be considered. The following list deals briefly with these difficulties.

1. *The costs and ease of measuring pollution must be considered.* Pollution must be monitored in order to determine whether standards are being exceeded. Noise from motor vehicles is difficult to control because of the numerous metering facilities which are required for monitoring.
2. *It must be possible to identify the source of pollution*, a task which can involve considerable difficulty and expense in the case of illegal discharge of oil at sea (Ch. 4).
3. *Ambiguity of the regulations may lead to costly court cases.* Often exceptions to rigid standards are necessary to avoid inequities or unwarranted costs. However, these exceptions can lead to greater ambiguity and hence greater costs of enforcement.
4. *Standards may be easier to enforce if they are supported by popular*

consensus. Lack of public support or resistance by political and economic interests sometimes lead to low levels of compliance and ineffective policing of the standards.

5. *The number and size of polluters may affect the ease of regulation.* As we shall see in Chapter 8 the problems of enforcing many mobile demolition firms working with asbestos products is very different from those regulating a few large factories manufacturing asbestos products.
6. *The cost of effective enforcement may vary considerably according to the nature of the pollutant.* The Royal Commission on Environmental Pollution (1976a), for example, recommended that enforcement of industrial air pollution regulations should continue to be the responsibility of the Alkali Inspectorate because local authorities could not be expected to have the required expertise for enforcement. Enforcement costs also vary according to the strategy adopted. Thus environmental quality objectives require more expertise than the uniform emission standard approach (Gardiner 1980). The licensing system, which is a flexible system of control, allows stringent control or relaxed controls when appropriate. However, to work effectively, it requires an authority capable of estimating damages caused by a discharge, competent to monitor discharges, and also competent to balance conflicting interests. If this competence is not available, then the protection agency is likely to be in a weak bargaining position with the polluter. Bouveng (1980: 333) believes this was the case during the expansion during the 1970s of the Swedish National Environment Protection Board:

   > According to common bureaucratic tradition, however, the expansion of the EPB took place mainly at the junior level. The same tradition caught the senior officers in a network of Committees leaving the licensing matters almost entirely in the hands of new forces. This has led to cases concerning heavy process industries being handled by Civil Servants whose understanding of industrial affairs has not yet reached a desirable level.

7. *The size of penalty may influence enforcement.* The larger the penalty the greater the deterrent. However, at some level the penalty will be considered by the enforcers and courts to be unreasonable. This could lead to less court action and greater avoidance action by the polluter.
8. *Substitution possibilities may affect enforcement.* When substitution of products involves few extra costs to manufacturers then enforcement is more likely to be effective. For example, manufacturers have accepted some tighter standards on asbestos fibres when substitution has been practicable, but have strongly resisted tighter standards when this has not been the case (see Ch. 8).
9. *The number of regulatory agencies may lead to problems of co-ordinating enforcement activities.* In the US, for example, conflict

between state and federal authorities sometimes reduces the effectiveness of enforcement.

These features of enforcement and characteristics of the regulatory approach can be illustrated by the operation of water pollution controls in the US and Britain, and industrial air pollution control in Britain.

The American permit system

Under the Federal Water Pollution Control Act Amendments of 1972, permits are issued by the Environment Protection Agency (EPA) or by the State. By November 1975, twenty-seven states had taken over the permit-issuing function, but the EPA still retained the right to determine the nature of the permit (Council on Environmental Quality 1976). Permit-setting takes place within a system of guidelines. By the end of 1974 effluent guidelines had been set for twenty-eight groups identified in the 1972 Amendments, and by 1 July 1975 over 40,000 permits had been issued. These permits last for five years and give details of the pollution allowed. The system is also supported by subsidies which reduce capital expenditure on treatment plants by between 55 and 75 per cent. Enforcement is carried out by the State, but can also be initiated by the federal government or by individuals through citizen suits. Heavy fines of $2,500 to $25,000 per day for the first offence and up to $50,000 per day for subsequent convictions provide a substantial sanction on non-compliance. The American permit system is subject to significant delays as a consequence of legal and administrative challenges to guidelines and to permits allowable within the political framework of the American constitution. In the first four years following the Federal Water Pollution Control Act Amendments, more than 10 per cent of major industrial permits issued by the EPA had been subjected to administrative hearings. In the first two years of operation there had been 150 lawsuits, mainly brought by industry, concerned with the establishment of effluent guidelines (Council on Environmental Quality 1975 and 1976).

Enforcement of standards has been far from satisfactory. The 1972 Federal Water Pollution Control Act Amendments established 1 July 1977 as the deadline for all publicly-owned sewage-treatment works to be using secondary treatment and all industrial plants to be using 'best practical' pollution-control technology. By this date only about 40 per cent of the 'major' municipal dischargers were in compliance, but 80 per cent of the 'major' industry dischargers were in compliance. In the optimism of the early 1970s the 1972 Amendments required that 'best available technologies economically achievable' were to be installed by 1 July 1983. The Clean Water Act of 1977 extended the deadline by one year and also made a distinction between toxic and non-conventional pollutants on the one hand and conventional pollutants on the other hand. Standards for the latter were to be based on guidelines which would take different cost considerations into account. In other words, standard-setting was to take into consideration both the potential damage and the costs of a pollutant, as well as the costs of abatement (Council on Environmental Quality 1979).

The British consent system

In England and Wales, a system of consent conditions was introduced in the 1951 and 1961 Rivers (Prevention of Pollution) Acts. As with American permits, limits are set on the quality and quantity of effluents. Less formal guidelines exist, much reliance being placed upon the Royal Commission on Sewage Disposal (1912) standards of 30 mg/l suspended solids and 20 mg/l BOD. Regional water authorities have in recent years been revising consent conditions to take into consideration local demands upon rivers, as well as environmental and economic circumstances. Following reorganization of the water industry, water authorities became responsible for standard-setting and enforcement, but they also became responsible for sewage works, and water supply. This change was influenced by the view that it would allow water authorities to concentrate expenditure where they could produce most benefits. The lack of an independent regulatory authority has, however, weakened enforcement policy, as the water authorities have become the main offenders through sewage works failing to comply with consent conditions originally imposed by river authorities before 1974 (Sandbach 1977a and b). Low levels of fines have also weakened the sanctions on non-compliance. Unlike in the US, initiation of legal action by members of the public has not been possible. Part II of the Control of Pollution Act 1974 would have established a more rigorous prosecution policy by allowing private prosecutions. The public would also have had a right to comment on consent applications and much more information on the regulatory process would have been made available for public scrutiny. Unhappily, restrictions on public expenditure have prevented implementation of these features of the Act.

A persistent criticism of the regulatory approach in Britain has been the secrecy of administration. While the public has often been uninformed, industry has more often than not enjoyed a close relationship with the regulators and enforcers of standards. On the other hand it has also been argued that greater fines and a more coercive approach, as in the US, has led to prolonged litigation, the resolution of some cases being delayed for up to ten years (Freeman III 1971). Storey (1977) has also suggested that a co-operative approach allows the regulator to gain access to more information and to accumulate sufficient expertise to be able to act in a positive constructive manner. Industry naturally wishes to protect trade secrets and most countries have legislation which prevents their disclosure. However, secrecy of information has in some countries, notably in the UK, given rise to serious problems of enforcement. The Rivers Prevention of Pollution Act 1961 made it a criminal offence for the water authorities to disclose information on the imposition of conditions or to disclose information on pollution which had been obtained either in connection with an application for consent or in the course of inspection, unless the disclosure was made with the consent of the person supplying the information or producing the pollution. It is no wonder this Act has been called a polluter's charter.

The Control of Pollution Act 1974, when fully implemented, will go a long way to remedying this problem of secrecy in pollution control. It will even allow individual citizens the right to initiate prosecutions when water pollution exceeds consent conditions. Unfortunately, this radical departure in policy, which resulted from public agitation in the late 1960s and early 1970s, has been delayed, perhaps indefinitely, while the water authorities struggle to set enforceable standards.

Industrial air pollution control in Britain

Responsibility for industrial air pollution is based in part upon statutory limits governing emissions from alkali/hydrochloric acid works and sulphuric acid works. Standard-setting for other processes is delegated to the Alkali and Clean Air Inspectorate which must ensure that industry and commerce adopt the 'best practicable means'.

The expression 'best practicable means' was incorporated in the Alkali Act of 1874. It enabled the Alkali Inspectorate to determine the best way of controlling air pollution from those works scheduled under the Alkali Acts. The only statutory definition of 'best practicable means' which is accepted by the Inspectorate, can be found in the Clean Air Act of 1956: '"practicable" means reasonably practicable having regard amongst other things to local conditions and circumstances, to the financial implications and to the current state of technology'. The assessment of 'best practicable means' is according to Damon, a former Chief Inspector, a compromise between: 1. the natural desire of the public to enjoy air; 2. the desire of manufacturers to avoid unremunerative expenses which would reduce their competitiveness; and 3. overriding national interest (Frankel 1974: 8).

Presumptive limits for emissions, based upon best practicable means, are laid down for various industrial and other emissions from scheduled plants. However, new works may be required to keep emissions below the presumptive limits. These presumptive limits and the failure to comply with them may be used in evidence in legal proceedings. In practice universal standards are often enforced instead of those best-suited to meeting the needs of local circumstances. Frankel (1974: 9) commented:

> The Alkali Inspectorate pays relatively little attention to local

conditions in its interpretation of 'practicable'. Occasionally (for cement works or lead works, for example) a large works must meet a stricter standard than a small works. But in general a single emissions standard applies to each class of works wherever the works is situated and whatever its size. Only the height of the chimney, used to dispense the emissions, varies with local conditions.

The Alkali and Clean Air Inspectorate is responsible for approving plant design and for regular testing and monitoring of emissions from scheduled plants. The number of works registered under the Acts at the end of 1977 was 2,150, involving the operation of 3,032 processes (*Industrial Air Pollution* 1977 (1979)). These works are responsible for three-quarters of the country's fuel consumption and represent the most persistent and troublesome sources of air pollution, from a wide range of industries including cement works, oil refineries, iron and steel, lead, power stations, aluminium smelters and ammonia plants, as well as the three works still registered for the alkali (saltcake) process. In 1975 only about 6,500 tonnes of such salts were produced, which compares with the 650,000 tonnes yearly in the days of the Leblanc process in the 1880s which had caused the Alkali Inspectorate to be formed in 1863. So although the official title of HM Alkali and Clean Air Inspectorate is in current parlance, most of the scheduled processes are no longer in the alkali industry. Non-scheduled works and all domestic pollution are controlled by local authorities under the Clean Air Acts of 1956 and 1968.

The Inspectorate works closely with industry and trade associations to encourage innovations and to keep a watching brief. They regard themselves as a team of 'trouble shooters' dealing only with technically difficult problems of control, and they try to find practicable answers to unresolved problems. Presumptive limits for emissions are reviewed by the Inspectorate from time to time to take into account improving technology. For example, the emission standards for cement works have been reduced from 0.4 grains/ft^3 in 1950 to a sliding scale of 0.2 to 0.1 grains/ft^3 (depending on the site rate of production) in 1975 (*Industrial Air Pollution* 1975 (1977)). The flexibility of changing presumptive standards without resort to Parliamentary legislation has advantages. However, these get changed only once every ten to fifteen years, and the statutory standards for alkali/ hydrochloric acid and sulphuric acid works have not been changed since 1906. Flexibility in operating the 'best practicable means' also governs the timing of introduction of new pollution control technology. After introducing pollution control technology, the plant is allowed to operate for its economic life, which assuming satisfactory performance will usually be for about ten years, before new demands for more stringent standards are made. This policy has obvious economic advantages over arbitrary quality standards which might enforce the adoption of new control technology at great cost shortly after recent best-known technology had been applied. Extension of deadlines and patching up of older equipment have also been allowed to prolong the economic life of pollution control technology.

Under the 1906 Alkali Act, prosecutions could only be brought by an alkali inspector with the consent of the Secretary of State. This power has rarely been used, and inspectors, in contrast to the US approach, have long preferred persuasion to prosecution, reserving the latter for severe and unreasonable infractions. Only when a firm's attitude is uncooperative, or is thought to be deliberately flouting the law, is prosecution felt to be justified. Most infractions arise from plant breakdown, but not all infractions are recorded, nor is the term itself without ambiguity. The Chief Inspector in 1951 admitted in his annual report that there was 'no well-defined dividing line between legal and illegal operation and a decision as to whether a given set of conditions shall or shall not be treated as an infraction (i.e. a contravention of the Alkali Act) rests, to a great extent, with the District Inspector' (Frankel 1974: 16). Thus the recording of infractions is somewhat arbitrary. For example, after W. A. Damon, Chief Inspector since 1929, retired in 1955 the high rate of infractions suddenly dropped. Ninety-three infractions had been recorded in 1954 but by 1958 these had fallen to twelve. The new Chief Inspector acknowledged that this figure did not represent an improvement in working conditions but rather that fewer visits were made to older works than in normal years. During this year an alkali order tripled the number of scheduled works under the Inspectorate's control (Frankel 1974: 17–18).

Between 1920 and 1966, when the Alkali Inspectorate was largely out of the public eye, there were only two prosecutions. Between 1970 and 1975, during a period of intense public interest, there were twenty prosecutions (Table 6.1). Infractions in the metal recovery works have been the most frequent with twenty-one infractions in 1975. Nine out of twelve of the prosecutions between 1967 and 1972 were also against metal recovery works (Frankel 1974: 19). The Inspectorate's justification for so few prosecutions is that legal action 'will not solve technical problems or breakdowns, teething troubles or the occasional results of industrial action' (Department of the Environment 1973).

Another sanction available to the Inspectorate is to refuse reregistration of a scheduled work, which must take place every year, but this is rarely contemplated. The Health and Safety at Work Act 1974 has made it possible to apply two new methods of enforcement: through prohibition and enforcement notices. Prohibition notices can take effect immediately or can be deferred. They would only be used if it became apparent that there was a serious risk to people. An enforcement notice can only be issued when a statutory contravention, i.e. failure to use the 'best practicable means', has occurred. These new powers have not, however, changed the Inspectorate's basic policy of trying to persuade rather than coerce industry into controlling pollution (Jones 1977).

Table 6.1 Alkali Act prosecutions

Year	Infractions	Successful prosecutions	Prosecutions % of infractions
1970	25	2	8
1971	38	2	8
1972	58	3	5
1973	59	4	9
1974	57	2	4
1975	70	2	3
1976	66	8	12
1977	91	8	9
1978	100	13	13

Note: Infractions and prosecutions are listed in the year in which they occurred, although a prosecution may relate to an infraction which took place the previous year. Figures taken from: Royal Commission on Environmental Pollution; 1976a: 63; *Industrial Air Pollution 1975*, Health & Safety Executive (1977: 5); *Industrial Air Pollution 1976*, Health & Safety Executive (1978: 4); *Industrial Air Pollution 1977*, Health & Safety Executive (1979: 2); *Industrial Air Pollution 1978*, (1980: 3).

In principle the 'best practicable means' offers an acceptable means of pollution control, as its operation should take into consideration economic, social and technical factors. As far as the Alkali Inspectorate is concerned any failure to improve air pollution can be attributed to what the country can afford and not to any failure in the administration *per se*. In the 106th Annual Report of 1969 it was made clear that the technology was available to ensure a clean and healthy environment but that the main constraint was economic. One year later the Chief Inspector against explained (Bugler 1972: 24):

> If money were unlimited there would be few, if any, problems of air pollution control which would not be solved fairly quickly. We have the technical knowledge to absorb gases, arrest grit, dust and fumes and prevent smoke formation. The only reason why we still permit the escape of those pollutants is because economics are an important part of the word 'practicable'. Most of our problems are cheque book rather than technical.

There is little doubt that the flexibility of the 'best practicable means' has distinct advantages, and it is less likely that its operation would result in greater costs than benefits as has been suggested for American air pollution control policy. Moreover, the system has helped considerably in reducing the problems of air pollution. It is for these reasons that the Royal Commission on Environmental Pollution and Government have reaffirmed their faith in the operation of the principle (Howell 1976).

Despite its professed economic and theoretical advantages, the operation of the 'best practicable means' principle has been much criticized. First it is claimed there have been too few inspectors adequately to monitor and regulate the scheduled processes. In England and Wales during 1972 there were only thirty-five inspectors responsible for dealing with some 2,170 registered works. As there were only fifteen administrative districts some of the inspectors were based as much as 240 km away from a plant under their control. The alkali inspector in Stafford controlled an area of roughly 8,000 km^2. This made it difficult to keep an eye on the behaviour of firms who might well avoid detection by discharging more harmful effluents at night, and although inspectors made 'spot checks' these have been considered inadequate: 2,500 'spot checks' were carried out in 1972 for 3,274 separate processes (Frankel 1974). By 1977 the situation, despite publicity, had changed little. Forty-two inspectors and four technical assistants still covered fifteen administrative areas containing 2,150 registered works. The total number of visits and inspections during 1977 was 15,745, representing on average some six and a half visits a year to each works (*Industrial Air Pollution 1977* (1979)).

Secondly, since the first Alkali Act of 1863, the Inspectorate has worked very closely with industry. Not only are inspectors usually recruited after many years of experience, but the underlying philosophy of the Inspectorate has been one of gentle persuasion. In this they are not really very different from other British inspectorates that have traditionally relied upon education, supporting advice and persuasion rather than prosecution (Hartley 1972). Critics have suggested that this relationship has been too close. Most firms are allowed to do their own monitoring with little independent assessment. While there has been a friendly relationship with industry, the same has not, it is argued, been the case with the public.

One case of close co-operation with industry and scant regard to public concern was the building of a Rio Tinto Zinc smelter on Anglesey. Standards of pollution control were estimated by industry and the Inspectorate, and given as evidence at the public inquiry prior to approval of the plant. These standards were then promptly relaxed with the Inspectorate's approval. West and Foot, who have described the controversy over the smelter, complain bitterly about the way the public were duped and about the Inspectorate's lowly opinion of public concern and planning procedure. Ireland, the Chief Inspector, reputedly dismissed the objections of one resident in the following manner (West and Foot 1975: 212):

> We do not accept that the estimates submitted by the Company at the public inquiry were binding in any way. They were given in all good faith as typical of what emissions were expected in order to meet our targets, on which the figures were based. They represented our preliminary estimate of what might be achieved on the evidence of known technology. The important point of the inquiry, in relation to air pollution, was that the company should meet requirements of the Alkali

Inspectorate. When we got down to details of design and in the light of practical tests carried out in the USA on full-scale plant, it soon became obvious that we had to change our original thoughts on prevention and dispersion in order to keep the project viable... Industry cannot be handicapped by rigid rules based on estimates.

It has been argued that the Inspectorate is reluctant to become involved in public debate, believing that the media have 'cruelly mis-used' information and would continue to do so (Department of the Environment 1971). Furthermore the Inspectorate believes that the public would be incapable of a sensible interpretation of the data. Annual reports of 1971 and 1974 refer to the poor image of the Inspectorate which it believed resulted from the unbalanced views of the media, the latter report lamenting that 'the work has become more onerous with the increased public interest in the environment...' (Department of the Environment 1974).

The Alkali Act 1906 made no requirement of secrecy in connection with the data collected, but on the whole the Inspectorate chose to do so by a strict (although unnecessary) interpretation of the Official Secrets Act of 1911. Despite the Royal Commission's protestations about secrecy in pollution control administration, the Health and Safety at Work Act 1974 (section 28) put a statutory restriction on the disclosure of information without the consent of the person by whom it was furnished. This makes it particularly difficult for the public to ascertain whether or not complaints have been acted upon. The Control of Pollution Act 1974 empowered local authorities, at their discretion, to collect and publish information on discharges. However, local authorities are unable to obtain information that is not currently being supplied to the Inspectorate.

The Inspectorate have defended their position *vis-à-vis* communication by arguing that they have co-operated closely with local authorities. Since 1958 inspectors have formally visited public health departments of local authorities in which registered works are sited at least twice annually. It has also been their increasing practice to set up local liaison committees. In 1977 there were nearly fifty of these committees in England and Wales. They are composed of public, local authority, industrial and Inspectorate representatives. This forum is meant to allow the public to voice their complaints, meetings being held from two to six times a year. Nevertheless there is nothing approaching the open system of government as found in the US. Ireland (1977: 8) defended this situation in terms of administrative efficiency:

> It has been said that, as Government servants, we should publicise everything about our testing and inspectings of all registered works, submitting our reports to local authorities. What an enormous exercise and, in my view, what a waste of expert professional time... I believe that when experts are employed to do a highly technical job, they should be allowed freedom to get on with it with the minimum of

distraction and with the greatest efficiency... We have moved steadily to a more open policy, but there are limits.

This elitist and technocratic stance has understandably led to public confusion and resentment. Most of the complaints from the public are channelled through the local authority health department, as few people know of the Alkali Inspectorate's existence. The anonymity of the Inspectorate is reflected in the protest campaign against a carbon black works in the Port Tennant area of Swansea. The United Kingdom Chemical Works (now UCB) opened its Port Tennant factory in 1948 and was registered under the Alkali Inspectorate in 1953. During the 1950s Port Tennant residents protested to the factory about pollution and petitioned the local authority and their MP. Another petition was presented to the Swansea East MP in 1966, but it was not until 1970 that the residents started to take militant action by blockading the works and forming an anti-pollution association. During the next three years much publicity in the press and on television followed the conflict between industry and community. During this long campaign, it was not until the first blockade in 1970 that the protesters learnt that the Alkali Inspectorate was responsible for regulating UCBs emissions (Hall 1976). Bugler commented on the anonymity of the Inspectorate: 'in no telephone directory in the country can you find a reference to the Inspectorate; inspectors can be contacted only if you happen to know their names, private addresses and telephone numbers' (1972: 8).

The final area of criticism has been the lack of definition, which we have noted, of what constitutes legal and illegal operation. According to Frankel the solution to this and the other problems was to transfer the responsibilities of the Alkali Inspectorate to the local authorities, which would in turn be more accountable, through direct elections, to the local communities affected by pollution. This solution was, however, rejected in the Fifth Report of the Royal Commission which was asked specifically to look at this issue. It was argued that the expert skills and competence of the Inspectorate should be preserved, as they were necessary for effective control (1976a: 24). The local authorities did not have the technical knowledge and experience required to exercise control over technologically complex processes, whereas

> The Inspectorate collaborate closely with industry in seeking solutions to pollution problems. The solutions which they eventually impose can be both tougher and more practicable as a result of this involvement; the technical expertise of the Inspectorate is both essential to, and fostered by, this collaboration. Moreover, because of their understanding of the industrial processes involved and the possibilities for pollution abatement the Alkali Inspectorate are able to advise industry during the design stage for new plants, so that pollution control requirements are taken into account from the outset.

The Royal Commission reaffirmed the effectiveness of the best practicable means principle and wanted it applied to all forms of pollution arising from a given technological process, and not as at present, only to air pollution control. The report recommended that there should be a unified Inspectorate (HM Pollution Inspectorate) responsible to the Secretary of State for the Environment. It would be responsible for dealing with all the technologically sophisticated problems of pollution control. The logic of a unified inspectorate according to Flowers (1977: 7), Chairman of the Royal Commission, was the belief that 'the ultimate source of pollution is a process which in general pollutes water and land as well as air, and that pollution can be transferred from one form to another by modification to the process and to the treatment given its various discharges'.

The main recommendation for tightening up the 'best practicable means' approach has been the suggestion that when a works is registered, consent conditions would be set which had to be complied with. This type of registration would be renewed every two or three years as long as there was not repeated infringement of the conditions. Power to refuse reregistration would rest with the Secretary of State. This change would give greater clarity and would make prosecutions for infractions easier, but the Royal Commission did not feel that a more vigorous prosecution policy was desirable. However, any breach of air pollution control requirements, should, in its opinion, be made public.

Conclusions

In this chapter we have reviewed the main arguments for and against the use of pollution taxes as opposed to a regulatory approach based on the use of pollution standards. The arguments in this debate are mainly hypothetical and abstract as there has only been a very limited experience of pollution taxes in practice. However, the advantages and disadvantages of pollution taxes, permits, consents and the best practicable means principle cannot be judged merely from a theoretical position in an ideal world.

There has been a tendency amongst the promoters of pollution taxes to assume that they can be implemented in an efficient way without distortion from political and economic interests. These interests have clearly affected the regulatory approach based on standards. Nevertheless the adoption of standards and their enforcement makes some use of economic principles in balancing costs and benefits of control. The limited experience of pollution taxes in operation has tended to show that they are just as much prone to interests seeking delays in implementing taxes at an effective level. The operation of different methods of control in practice is constrained by institutional arrangements, pressure groups and economic interests. It is these features of pollution control that we must now consider in greater detail.

CHAPTER 7

Public participation and economic interests in pollution control

An adequate explanation of pollution control policy must take into consideration not only information on dose-effects, cost-benefits, and different styles of administration which structure decision-making, but also social and economic interests. To explain the development of law and its enforcement it is necessary to consider the relevance of public opinion, economic interests and the contribution of pressure groups and individual personalities. Accounts of decision-making or non decision-making depend upon different theoretical and ideological assumptions. In this chapter our attention will focus upon varying assumptions about the role of pressure groups, elites and other interests in decision-making. A start will be made by looking at the behavioural/pluralistic perspective which tends still to dominate political science.

The pluralist perspective

Pluralists assume that the political system is composed of a multitude of competing groups who strive to influence decision-making in their own favour. There is no absolute consensus upon the nature of a social problem and how it should be resolved, but while there is conflict between different groups it is of limited extent because there can be no dominant ruling class able to suppress a subordinate class.

The pluralists claim that through electoral reform and the growth of State activity there has been a transformation of society away from that of the mid-nineteenth century, which was more dominated by unconstrained capitalists. There has, it is argued, been a diffusion of power among a wider range of groups. In this sense the pluralist perspective could be claimed to be historically specific, the extent of pluralism being a relative notion. It may therefore be regarded as a state of affairs to be tested as well as a state of affairs to be striven for. The view that there is class domination by a minority with a vested economic interest is rejected by pluralists, who claim that any attempt at such domination would be checked by a majority through elections or various other constitutional and legal checks on the abuse of power. Because of these checks, the pluralist perspective, while accepting different interests which struggle for power and influence, falls back upon the belief that any social change must be acceptable to the majority of interest groups and people involved – hence the importance which pluralism attaches to public opinion (as reflected by the media, pressure groups, civil servants and experts) in the development of legislation and policy.

In the pluralist democracy then, public opinion or elites of concerned individuals, can recognize the danger of a pollution problem and bring about the appropriate legislative and/or administrative change. Ashby (1976: 724) commented:

> ... Legislators rarely lead public opinion, they have to follow it, or at best anticipate it by a very short head. But there is a subtle two-way influence. Sometimes, as over the disposal of toxic wastes, politicians are driven to and by public opinion. At other times, as over the Health and Safety at Work Act, the politicians try to educate public opinion: laws can be pedagogical as well as prescriptive. So the politician has to try to interpret social norms as well as to influence them; and this is both difficult and capricious.

Given the dependence of the pluralist perspective on the influence of public opinion, the question arises as to how public opinion can change in such a way that it brings about a political response to an issue. To answer this it is necessary for the pluralist to evoke a number of exogenous factors. Certainly public opinion is a response in part to the 'facts' of the particular problem: the extent of pollution, its imageability in terms of dramatic consequences, its economic costs, and the availability of practicable remedies are important. These facts are, however, only prerequisites. What actually sparks off a rise in public concern may be a 'chance' event – the London smog of 1952, the *Torrey Canyon*, Santa Barbara and *Amoco Cadiz* oil spills, for example. Such events capture the attention of the public and reveal an isssue that needs to be tackled (Downs 1972). In the case of illegal disposal of toxic wastes, the problem was well known to the Department of Housing and Local Government in 1964, for it set up a committee to investigate it. But it was not until 1971 that the press reported that illegal dumping of drums containing cyanide and other toxic wastes was being carried out by lorry drivers who received bonuses. This scandal precipitated government action, and the law was tightened up by the Deposit of Poisonous Wastes Act 1972 a full two years before the comprehensive Control of Pollution Act which also covered the disposal of toxic wastes (Ashby 1976).

The reason for the broader national and international concern about pollution problems during the 1960s and early 1970s are only partly explained by reference to the pollution problems themselves and to the events which are associated with a 'take-off' in public concern. Durkheim's (1950) view that all societies have social problems and that these must compete for public attention helps to explain the rise of public concern about pollution. As competing social problems such as poverty, housing and racial tension became less serious in the affluent post Second World War period, then pollution problems (as low-order social problems) were able to capture public attention.

An underlying feature of the pluralist perspective is the requirement that pollution problems shall be dealt with in open debate with a range of interest groups seeking to influence public opinion by advocacy and persuasion. The subject matter of the pluralist study is therefore the behaviour of individuals, groups and organizations. The focus of attention is upon political actors and the persuasiveness of the various conflicting arguments. The pluralist account and its limitations can best be demonstrated by reference to case studies. The history of the origins of clean air legislation in Britain, it will be claimed, is accounted for more or less satisfactorily in different ways according to whether one views events in terms of political actions or economic considerations. Likewise the early history of industrial air pollution control is used to demonstrate the part played by political interests. The importance of economic interests as opposed to the behaviour of a plurality of interests can be further demonstrated in more recent pollution issues.

The historical roots of the British Clean Air Act 1956

Ashby and Anderson (1976 and 1977), in an impressive and extensive study of the development of the clean air issue in Britain, illustrated some of the features of the pluralist/behaviourist mode of explanation. Their account is sophisticated and they recognize the importance of an historical account in order to explain why there was clean-air legislation in the 1950s rather than at any earlier period. In this respect they offer an account that demands more serious consideration than the historical accounts often provided by the political scientist. For example, Sanderson's (1974) analysis of the Clean Air Act 1956 rests solely on the activity of the National Smoke Abatement Society, the public opinion resulting from the London smog of 1952, the Beaver Committee reports, and the initiative of Gerald Nabarro, an MP.

Ashby and Anderson's account is based largely upon an almost blow-by-blow reconstruction of the action taken by the political participants during a period of over a hundred years. Pressure groups, legislators, industrialists, administrators, and scientific experts are locked in combat, but in the final analysis, as Ashby (1978) commented elsewhere, legislation was the outcome of 'a glacial movement of public opinion towards a consensus to outlaw smoke'. Pollution control must be technically feasible and enforceable, but nevertheless public opinion is the essential element in the explanation. Pressure groups are important promoters of public opinion, but once again there has to be a wider consensus if such public opinion is to become established.

As early as 1842 there was a pressure group, the Manchester Association for the Prevention of Smoke, which is seen as a precursor of later pressure groups. Individuals such as Mackinnon, a member of the landed gentry, are also credited with instigating public opinion. During the decade 1841–50 he was instrumental in promoting six smoke-abatement Bills. Although none was enacted, Ashby and Anderson (1976: 287) suggested that Mackinnon was responsible for increasing public interest in clean air: 'his years of patient and urbane pressure for control of smoke had raised a groundswell of public opinion which carried the campaign into the next decade. For the first time, the problem of smoke abatement had been subjected to most vigorous public scrutiny. In the more favourable political environment of the 1850s, Mackinnon's cultivation of the public conscience bore fruit'.

The story is continued with Mackinnon being supported in his campaign by John Simon and Palmerston. Counter-publicity comes from industrialists, especially the renowned John Bright who is better known for his part in the campaign against the Corn Laws. Nevertheless, public opinion and the more prosperous economic circumstances allowed compromise legislation in the form of the Smoke Nuisance Abatement (Metropolis) Act 1853, which resulted in limited control of industrial smoke in London.

After 1853 public interest was 'diverted to international events and away from such domestic issues as smoke', although local authorities other than London were empowered to control industrial smoke under various 'smoke clauses' in the Sanitary Acts of 1858 and 1866 and the Public Health Act 1875. Ashby and Anderson assess this local legislation in terms of three limitations. First, the law was often either weak or unenforceable. Secondly, local authorities were unwilling to adopt a strong enforcement policy for fear of discouraging industry. Thirdly, the little that was done in controlling industrial smoke was more than dwarfed by the increase in domestic smoke.

As in other public health campaigns the scientist is given major credit for fostering public concern. In the case of clean air it was Russel's investigations of London fogs of 1873 and 1880 which gained the attention of the historians. Russel helped to link the fatalities from fogs and the damage to property from smoke with the domestic burning of coal. It was in this climate of concern that a new campaign for clean air was begun by the Fog and Smoke Committee, which was formed in 1880 under the joint auspices of the National Health and Kyrle Societies. This new campaign committee helped to mobilize experts in fuel technology. By this time Ashby and Anderson claimed that the technology for controlling smoke was available and economically viable. The failure to control smoke was accordingly attributed to social values (Ashby and Anderson 1977: 14):

> After a year of intense activity it must have become apparent to the Fog and Smoke Committee that the most intractable problems in the case of domestic smoke were not technology or economics: they were in sociology. There were already stoves on the market which would burn smokeless anthracite coal or coke. Anthracite coal cost no more than ordinary coal. But the householder, eager enough by now to suppress smoke from factory chimneys, was not willing to forego his own cheerful fireplace for the silent and colourless emanations from a closed stove; nor, by all accounts, were servants willing to exchange the daily chore of cleaning and lightning fires for the more skilled, but certainly cleaner, job of tending an anthracite stove.

So from 1880 until the Clean Air Act 1956, the major constraint was one of public attitudes and not economic circumstances. Their argument here seems to derive largely from several references to contemporary claims that closed stoves burning anthracite would prevent smoke pollution, but Englishmen did not like closed stoves. They support this argument with the fact that there was very little political discussion concerning the economic implications of smoke control when Lord Stratheden and Campbell introduced a Smoke Nuisance Abatement (Metropolis) Bill on ten occasions between 1884 and 1892. However, they do point out that the Prime Minister's opposition to the tenth Bill was partly based upon economic considerations. If anthracite were to be introduced in any quantity, the Prime Minister stated, it would 'enormously... increase the price of that coal' (Ashby and Anderson 1977: 2). The remaining part of Ashby and Anderson's work concentrates therefore, on the activity of various political participants and events which were deemed to influence public opinion.

Ashby and Anderson argued that after the turn of the century the smoke problem became less serious. One of the principal reasons they offer for this was that gas increasingly became an economically viable alternative to coal for domestic and industrial consumption (Fig. 7.1). The decrease in smoke fits in nicely with their view that domestic smoke control was technologically and economically viable. They then give further evidence for the gradual change in public opinion in favour of smoke control. Six more Bills aimed at controlling smoke were introduced between 1922 and 1926 but these all failed to materialize. The Coal Smoke Abatement Society, formed after the exhibition at south Kensington, and the Smoke Abatement League joined together in 1929 to form the National Smoke Abatement Society, which was to carry on the campaign. They suggested that an increase in awareness of the clean air issue was demonstrated towards the end of the 1930s when the coal merchant's lobby began to rec-

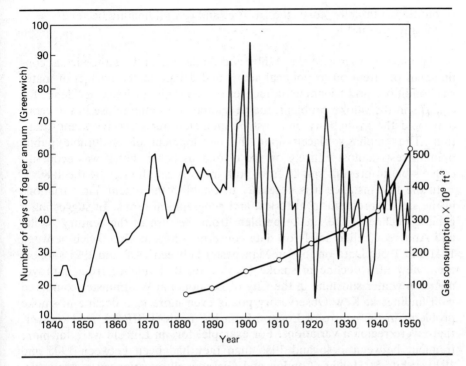

Fig. 7.1 Fogs at Greenwich. Number of days of fog per annum during day at Greenwich, 1841–1950, compared to sales of gas by statutory undertakings in Gt Britain, 1882–1950. *Source*: Ashby and Anderson (1977: 194).

ognize the economic threat, and in 1935 the Coal Utilization Council held a convention to promote the use of coal.

The London smog of 1952 has often been considered a crucial factor in the development of clean air legislation. Some 4,000 extra deaths have been attributed to the disastrous fog which lasted from 5 to 9 December. However, Ashby and Anderson pointed out that there were several nasty fogs in the 1870s and 1880s. One London fog in 1886 raised the mortality rate by 40 per 1,000 to the level of the great cholera year. This was at a time when they claimed that smoke could have been prevented. The essential factor which they emphasized in explaining why clean air legislation had to wait for seventy years was the state of public opinion. While most attention is devoted to the role of political participants, Ashby and Anderson (1977: 24) also offered another explanation:

> If one then asks why a fog in 1952 could precipitate legislation which the fogs of the 1870s and 1880s failed to precipitate, one plausible answer is the depletion of domestic servants. When the householder himself or his wife has to clean the grate and lay the fire, devices like the continuously

burning anthracite stove, the gas fire, and central heating become socially acceptable.

The main weakness of the Ashby and Anderson study is the virtual confinement of attention to political actors and debate to the neglect of material conditions and economic influences on the clean air issue. Evidence for changes in the smoke problem, the development of energy use in the economy, and the availability and cost of smokeless fuels receive scant attention. The emphasis placed upon the development of favourable public opinion rests almost entirely on the erroneous belief that it was economically viable to prevent smoke from both industry and houses by the 1880s.

In the first instance it is necessary to establish the extent and variations in the smoke problem in historical and geographical terms. In suggesting a general decline of the smoke problem from the turn of the century Ashby and Anderson relied solely on data concerning fogs at Greenwich between 1841 and 1950. Data on fogs in Manchester between 1892 and 1927 support their view of a decline in smoke (Taylor 1929). Evidence from improvements in winter sunshine in the City of London and Westminster compared with findings at Kew Observatory points even more to a decline in smoke problems in the centre of London from the 1880s (DSIR 1937). However, there were regional variations. For example, fogs in Lincoln were far more frequent between 1946 and 1956 than they had been between 1905 and 1920. While St Helens, London and Glasgow all experienced a steady decline in deposited matter in the thirty years up to 1944, at two sites in Leeds the amount of deposited matter increased during the same period (DSIR 1949). Later studies of smoke at four sites in Glasgow, Greenwich and Stoke-on-Trent demonstrated a 30 per cent decline in smoke from 1939–44 to 1944–9. There was a further decrease in smoke at eleven sites from 1944–9 to 1949–54.

The question then arises as to how one can account for historical and geographical variations in smoke. The rise of gas consumption, suggested by Ashby and Anderson, was only one factor. There were in fact three factors which can account for the changes in smoke concentration. These were: first, changes in the use of fuels, secondly the development of fuel economy, and thirdly changes in location of industry, commerce and population. Let us look at the relevance of these three factors in turn.

The consumption of coal is of critical importance. Table 7.1 shows coal consumption according to use in 1869, 1903 and 1953. Although gaps exist in the long-term trends, it is fairly clear that domestic consumption of coal rose steadily throughout the period 1869 to 1939 when it reached a peak of about 40 to 50 million tonnes per annum. By 1953 domestic consumption of coal had fallen to 37 million tonnes per annum. However, this figure is artificially low, as coal rationing had been in existence since 1942. Domestic coal consumption dropped from 42 million tonnes in 1943 to 40 million tonnes in 1944 and to 37 million tonnes in 1945. It then stabilized

Table 7.1 Consumption of coal in the UK according to use (million tonnes)

	1869	1903	1953
Gas works	6	15	28
Electricity	–	3	36
Railways	2	12	14
Domestic	19	32	37
Industry, etc.	68	105	90
Total consumption	95	167	205

Sources: Mitchell and Deane (1962); DSIR (1963).

at around this level through to 1958. The Ridley Report in 1952 suggested that domestic consumption would have been between two million and ten million tonnes a year higher (their own guess was five million tonnes) if the supplies had been available.

Most of the coal consumed in the domestic sector was burnt on the open fire where there had been little overall improvement in efficiency of combustion or smoke control. As a consequence domestic consumption of coal had by 1939 become the main contributor to the overall quantity of smoke, and was an even more important contributor to the nuisance created by smoke. The quantities of raw coal for domestic and industrial consumption in 1953 were 37 and 90 million tonnes per year respectively, and yet more smoke came from domestic consumption as domestic combustion was less efficient. The Beaver Committee Report (1954) estimated that in 1953 900,000 tonnes of smoke were created by burning domestic coal, 300,000 tonnes came from the railways and 800,000 tonnes came from industrial and miscellaneous sources. Moreover, although domestic smoke was less dense than that from industrial chimneys it was discharged at a low level and was consequently more harmful. The proportion of domestic smoke was also much higher during winter months when it caused most harm to health.

As domestic coal consumption declined between 1939 and 1944 it is likely that this made a contribution to the overall reduction in smoke. The less dramatic decrease in smoke concentrations from 1944–9 to 1949–54 is, however, more likely to have been a result of reductions in industrial smoke. The reason for this lies in the fact that the decrease in concentration of smoke during the summer months was only slightly less than during the winter months. As domestic fuel consumption is considerably higher in the cold winter months, the main contribution must have come from non-domestic sources (DSIR 1955).

Consumption of coal by industry shows a rather different trend. Growth in consumption of coal grew steadily between 1869 and 1903 but at a much slower rate than in the domestic sector. Between 1903 and 1953 there was a slight decline in industrial consumption of coal. Two factors can account for these changes: first the development of fuel economy and secondly the switch to coke, gas and electricity. The first factor is the more important consideration. From the beginning of the Industrial Revolution and throughout the nineteenth century there were massive strides in improving fuel economy in production. Jevons (1906) was particularly impressed by the economy of fuel in the motive power behind the great industrial expansion in textile and other industries.

While some of the increase in fuel economy resulted from making better use of heat produced from furnaces, fuel economy also derived from the more complete combustion of fuel. In a lecture to the Leeds Philosophical Society on 3 March 1896, Cohen (1896) observed that soot given off from factory chimneys was at the lowest estimate 0.5 per cent to 0.75 per cent of the coal, whereas the average for house coal was over 5 per cent. Littlejohn (1897), Medical Officer of Health, made a survey of industrial boilers in Sheffield in 1897 and found that 579 out of 1,116 boilers had some form of smoke-preventing appliance. Reduction of smoke in many industrial areas depended largely upon fuel economy. In 1929, three million tonnes of coal were burned annually in Manchester, of which only 750,000 tonnes were used for domestic purposes. Taylor (1929), Assistant Medical Officer of Health, claimed that the significant reduction in smoke in the thirty years prior to 1929 was caused principally by a 75 per cent reduction in industrial smoke emissions. Reduction in industrial smoke meant that for many large towns domestic smoke had become the principal component of the total smoke problem. In Glasgow, for the year ending 31 March 1929, the average weekday ratio of domestic smoke to industrial smoke was 3.59, on Saturdays it was 5.71 (see DSIR 1929).

Reductions in smoke in some areas resulted from a change from coal to electricity or gas. In the County Borough of Rochdale, reduction in smoke pollution from 2,500 to 1,800 tonnes per 100 km^2 per annum between 1916 and 1927 has been attributed in the main to the adoption of electricity by the textile mills (Anderson 1928). Despite the greater costs of electricity and gas the greater controllability led to economies in certain industries. The substitution of smokeless fuels for economic and practical reasons can be illustrated by reference to a study by Beaver (1964) of the pottery industry. Prior to 1927 the potteries were entirely dependent upon coal for firing clay. The Staffordshire potteries were renowned for their bottle-oven kilns belching out smoke. In 1927 and 1932 respectively the first electric and gas kilns were introduced. These were already beginning to have their impact by 1938 although 80–90 per cent of all ware was still fired in bottle-ovens by coal. By 1953, one year after the London smog, the percentage of biscuitware fired by coal had dropped to 50, and of glossed and decorated ware to 20 and 17 respectively. Nine years later coal was used for firing only 9 per cent of biscuit and only 1 per cent of glossed and decorated ware. Smoke control in the potteries owed little to the advent of legislation but was almost entirely dependent upon the availability of superior production processes that involved the use of smokeless fuels.

The third factor affecting smoke pollution was changes in location of industry, commerce and population. In the first study of sootfall in London, reported in *The Lancet* (6 Jan. 1912), it was noticeable how levels of pollution varied remarkably within and outside the metropolis. In the city areas there were 255 tonne/km^2 deposits per year, in the south-west district and outside the Metropolitan area in contrast there were 180 and 76 tonne/km^2 deposits per year respectively. Concentrations of housing and industry are consequently important factors in the local smoke pollution problem.

Throughout most of the nineteenth century there was not only rapid increase in size of urban areas but also a concentration of people and industry in the central areas. Suburban development, particularly from the 1880s, which had been facilitated by development of local railways, tramways and bus services, reduced the pressure on the central areas and in some cities allowed a movement of people and industry out of central areas.

Reduction in smoke in central London from the 1880s can be explained in part by this change in urban development. The night population of the City of London fell from 74,897 in 1871 to 50,526 in 1881, a decrease of 32.5 per cent, while the day population rose from 170,133 in 1866 to 261,001 in 1881, an increase of 53.4 per cent (Ashworth 1954). While the central area attracted more commerce many industries had begun to move out. According to the evidence of a surveyor to the Royal Commission on London Traffic in 1904, many employers had moved out of London; these included ten printing firms and several engineering firms (Brown 1959).

During the interwar period several changes helped to loosen the urban structure further. Changes in transportation, especially the development of a mass car market and the replacement of trams by motorbuses, facilitated suburban development. Formation of the Central Electricity Board in 1926 and development of the national grid system liberated industry from traditional areas, encouraged growth of industry and population in the more prosperous Home Counties, and Midlands, and also helped industry to move out of central areas into the suburbs. In a study completed in 1933 of industrialization in the north and west suburbs of London it was shown that over one-third of the industries had moved from the centre of London, less than a quarter of the industries had been established before 1918, and the remainder belonged to newly-established firms (Ashworth 1954). The population of the County of London reached a peak of 4,563,000 in 1901 and declined to 3,353,000 in 1951, while the population of Greater London rose from 6,586,000 to 8,348,000 during the same period. Variations in dispersion of population, industry and commerce are probably of great importance in accounting for regional variations in trends of smoke concentration. Movement out of congested central areas certainly helps to explain decline in smoke concentrations despite an overall increase in domestic smoke.

The decline in smoke pollution before the Clean Air Act 1956 can therefore be understood simply in terms of economic and urban changes without reference to changes in public opinion. It can be further argued that the constraints on adopting domestic smoke control legislation prior to 1956 were mainly of an economic nature rather than the result of unfavourable public opinion. Ashby and Anderson claim that after 1880 domestic smoke control was technically and economically viable. They suggest that the decline in domestic servants might have encouraged a switch away from coal. For the elite, no doubt, the use of electricity and gas for room heating and water heating had its attractions despite the greater cost. However, use of coke and smokeless solid fuels would have required much the same housework as the use of coal. On the whole, gas and electricity were used for cooking and lighting mainly where convenience and controllability made them desirable sources of power (Ridley Report 1952). In a Report on Smoke Abatement by the Royal Institute of British Architects (RIBA) in 1929 it was claimed that the cost of gas or air heating was about two and a half times, and the cost of electricity about seven and a half times, that of coke for continuous heating. Gas and electricity were therefore a luxury for the rich, and their contribution to heating the domestic home in 1952 was still of minor importance. The Beaver Committee also felt that gas, electricity and oil were unlikely to be of much contribution to the control of smoke in the immediate future.

Coke and solid smokeless fuels were the only serious contenders against coal. Before the Second World War the main problem with adopting solid smokeless fuels was the lack of suitable open grates – adequate combustion was only practicable in closed stoves or boiler plants. The capital costs of replacing old open grates was the main constraint on universal measures of smoke control. Ashby and Anderson's study ignores this vital factor of converting from coal to smokeless fuels. The implementation of clean air zones in Manchester prior to the Clean Air Act demonstrates this economic constraint. Powers for the zones were first obtained in 1946 and exercised in 1952. They covered areas of mainly offices, administrative premises, commercial buildings, shops and department stores. These business premises had already become predominantly dependent upon gas and electricity, and over half the solid fuel consumed was smokeless. Malcolm (1977: 8) explained why Manchester City Council were unable to extend the zones to private housing:

> For housing, finance was only available from the Government in connection with Housing Act 'improvement grants' for replacing old grates in existing dwellings. Since Council finances were organised to be in keeping with Government finance policy, rate and loan money could only be used for modernising appliances in a very limited range of

premises, such as churches. To create smokeless zones by decree in established housing areas was, therefore out of the question, assuming individuals would not be prepared en masse to pay the whole cost of conversion.

In Coventry, smokeless zones were also confined to commercial areas for the same reason. However, in Bolton the local authority did establish a clean air zone in 1954 which included both factories and domestic property, but they were forced to operate a grant scheme in order to persuade householders to undertake the necessary alteration costs. In economic terms then the reasons for the delay in implementing clean air legislation were first the cost and availability of alternative fuels, and secondly the cost of conversion.

Ashby and Anderson's study fails to take into consideration technical improvements to open fire-grates and openable stoves from the 1920s. Before this there was no cheap substitute for the old-fashioned open fire-grate. The rich may have been able to afford stoves, but such stoves would have been beyond the means of most people, and there would have been no financial advantage. From the 1920s, however, gas undertakings in particular had an economic interest in producing simple and improved ways of burning coke. In collaboration with solid fuel appliance manufacturers they developed simple open fire-grates which were capable of burning gas coke. The Second World War brought an added stimulus to the search for greater economy in home heating. Scarcity of coal and other fuels during the war and in the decade following were important in this respect. The Fuel Efficiency Committee was set up in 1941 and government reports on fuel policy, the Simon Report of 1946 and the Ridley Report of 1952, made recommendations for fuel efficiency which were in harmony with smoke control. Following the war much progress was made in developing improved open fires. The improved simple open fire-grate had a room-heating efficiency of 25 to 35 per cent with coal and 35 to 45 per cent with coke. This compares with 20 to 30 per cent for the old-fashioned open fire-grate (Beaver Committee 1954).

The Ministry of Fuel and Power were able to produce a list of recommended appliances which would lead to more efficient combustion and at the same time give rise to little smoke. From 1948 all new local authority houses had to install appliances selected from the list. The coke department of the Gas Council also had its list of suitable appliances for burning coke. In six years from 1947 about four million approved appliances were delivered to distributors in the home market (Beaver Committee 1953). By 1952 the total costs of using coke for room heating for new houses were similar to those of using coal, despite the slightly greater costs of appliances and their installation (Ridley Report 1952). In actual running costs coke was considerably cheaper than coal for room heating.

The Beaver Committee (1954) believed that coke would be the main solution to the smoke problem, despite short-term difficulties of supply, as the cost of heating with coke was appreciably less than that of other solid smokeless fuels and house coal. The Committee recognised that the possibility of increasing the supply of gas coke depended upon the availability of coal, upon the enterprise of the gas undertakings in widening their market for gas, and upon the extent to which oil or non-coking coal helped to release coke then used by other domestic consumers. However, the Committee recognized that the main problem was that of replacing the open fire now in use: 'a more serious difficulty is that coke – even the improved product which we envisage – cannot burn satisfactorily on the majority of open fires now in use. The provision of suitable smokeless fuels is therefore only one part of the solution; the other is the provision of suitable appliances' (1954: 24).

The post-war development of the improved open fire-grate was absolutely crucial to the control of smoke. In an historical study of the development of smokeless zones Malcolm (1977: 8) even considered that the provisions in the Manchester Corporation Act 1946 for smokeless zones were premature 'because Raven grates were not available and neither was fuel'. The Beaver Committee estimated that the cost of conversion would not normally exceed £10. However, even with the reduced cost of conversion to smokeless fuels in the post-war period, it did not think people could be compelled to purchase necessary appliances to burn smokeless fuels in smoke control areas without a grant. They suggested a 50 per cent Exchequer grant to be supplemented by a local authority grant. The willingness of the State to intervene through the 1956 Clean Air Act by providing conversion grants rested upon a better understanding of the financial costs and damage to health caused by smoke. The investigations of Atmospheric Pollution by the DSIR, the lobbying of the National Smoke Abatement Society, the London smog of 1952, and the Beaver Report all contributed to this better understanding. Moreover, greater State intervention in social welfare after the Second World War facilitated legislation involving State expenditure on grants.

Contrary to Ashby and Anderson's thesis, it has been argued here that adoption of domestic smokeless fuels in the 1880s was impracticable and the conditions only began to become favourable in the post-Second World War period. Consequently the role of gradually changing public opinion must be discounted as the determining factor in the development of clean air legislation. Growth of electricity and gas supplies was of only minor importance in domestic smoke control. However, the growing economic interest of gas undertakings in developing cheap and efficient appliances for burning coke, and the development of fuel economy policies during and after the Second World War were much more important factors. Scarrow (1972) has shown that implementation of the Clean Air Act 1956 itself was dependent upon the availability and price of fuel. Market factors alone accounted for considerable voluntary conversions. A survey in London in

1960 revealed that 76 per cent of householders had already converted to smokeless fuels or for other reasons failed to claim the grant due to them. The overall extent of voluntary conversions outside London was estimated to be 19 per cent.

Like earlier legislation aimed at controlling smoke from industry, the Clean Air Act was riddled with loopholes, which enabled industry to develop its own fuel policy almost independently of the Act. First there was a seven-year exemption period for converting to suitable fuels; and secondly the adoption of smoke-control areas (rather than smokeless zones) was the discretionary responsibility of local authorities. The price and availability of fuels and the interests of industry helped to determine the implementation of pollution control. Happily the economic advantages of oil, gas and electricity increased and in the 1960s became the prime motive power behind industry. Domestic smoke control continued to lag behind industrial smoke control for economic reasons. While in 1956 domestic smoke represented just over 60 per cent of the total smoke, by 1962 it represented well over 80 per cent of the total. So in the final analysis smoke control has been the outcome principally of economic factors involving the development of alternative fuels to coal. As an influence on smoke control legislation, the increase of public concern about smoke has been exaggerated.

The Alkali Act 1863

An historical illustration of how law accommodates economic interests and resolves conflict between strong interest groups is provided by the creation of the Alkali Act of 1863. Prior to the French Revolution, France obtained soda from the ashes of marine plants imported from Spain, Sicily, Tenerife and Britain. As supplies ceased during the Revolution, Le Blanc came up with a method of chemical manufacture. Soda manufacture in Britain, however, did not begin until 1823. This followed the withdrawal of a tax on common salt, a necessary material for the production of sodium carbonate which was used in the manufacture of soap, glass and textiles. Before the tax was withdrawn the price of salt was as high as £30 per tonne, but by 1861 was only £5 per tonne. Such was the early stimulus to the development of the chemical industry!

In the early days no attempt was made to control the hydrochloric acid fumes which were a consequence of manufacture.

In 1830 Gossage started a Le Blanc works at Stoke Prior in Worcestershire. Complaints were received from local landowners and farmers. In 1828 Muspratt was fined one shilling for causing a public nuisance in Liverpool, where his first works were sited. In 1831 he was indicted but acquitted. From time to time he had to pay off farmers neighbouring his works near St Helens.

In response to these and as a result of his own personal discomfort from the fumes he set about trying to prevent the nuisance, and in 1836 he was able to patent a technique of condensing the acid fumes. Hardie (1950: 17–18) described how Gossage came upon his invention:

> Near his alkali works stood a derelict windmill which he filled with gorse and brushwood from the surrounding countryside, irrigating this packed tower with a downward stream of water and introducing his muriatic acid (hydrochloric acid) at the top with the water. Gossage found that little or no acid fumes appeared at the bottom.

The process was refined with the acid gas percolating through a deep bed of coke, in small lumps and in a high tower. At the same time a supply of water flowed very slowly over the coke, which provided a large moisture area for gas absorption. The invention was widely adopted by other manufacturers. Muspratt's Vauxhall works in Liverpool, which employed 130–150 men, had eight condensing towers which produced 5,000–6,000 tons of hydrochloric acid a year. Despite this, in 1858 Muspratt was forced to close the works and move to Newton, where he had already established an alkali works, when the corporation was successful in obtaining an indictment.

It was at St Helens and Widnes that the alkali industry expanded. Muspratt became associated with St Helens, while Gossage, Deacon and Hutchinson developed the industry in Widnes from 1847. Widnes was ideally situated on the Sankey Navigation extension and the south-west Lancashire coal field. It grew to be the most important heavy chemical manufacturing town in the world. St Helens was also suitable especially as local coal had a larger than usual content of sulphur.

By 1861 there were fifty soda-manufacturing establishments, employing some 19,140 hands, with products valued at some £2.5 m. per annum. Concentration of industry made it more difficult for the victim to obtain damages or an injunction against the fumes as it was more difficult to prove from which of the many chimneys the acid fumes originated and spoiled the farmers' crops or woodlands. Between 1852 and 1862 the alkali trade had doubled. By this time the complaints about air pollution damage from farmers and landowners had grown, for although the technical means to control the hydrochloric acid fumes were in fairly general use they were often inefficiently applied. *The Times* (12.5.1862) described the consequences as follows: 'whole tracts of country, once as fertile as the fields of Devonshire, have been swept by deadly blight till they are barren as the shores of the Dead Sea'. Hutchinson of Widnes, in response to the landed interest, persuaded most fellow manufacturers voluntarily to 'police' each others' factories to control fumes, but this was insufficient to stifle complaints. When a group of protestors approached Lord Derby he was sympathetic to their complaints, as he also owned an estate blighted by the fumes. Lord Derby claimed that the annual loss to his own estate and those of the other landed gentry around St Helens was £200,000. In 1862 he was successful in his demand for a Lords' Select Committe 'to inquire into the injury from noxious vapours evolved in certain manufacturing processes, and into the state of the law relating to these'.

The Select Committee with Lord Derby as its chairman noted that a similar process to that of condensing hydrochloric acid was used in the manufacture of sulphuric acid, but the process was carried out more efficiently because sulphuric acid was of much greater commercial value. Hydrochloric acid, on the other hand, was used mainly in the production of bleaching powder and there was already four times as much hydrochloric acid as there was demand for. The bulk of unmarketable acid was simply thrown into the rivers. There was not therefore the same economic incentive to condense the acid, but none the less the threat of the landed interest was a real one. The landed interest was still a strong force in Parliament and although the industrialists had gained a victory for laissez-faire, with the repeal of the Corn Laws, they were still a force to be reckoned with. In such circumstances the alkali trade admitted the need for special legislation designed to control the acid fumes, while at the same time emphasizing the importance of the chemical industry to the economy. What the manufacturers wanted was to reduce the threat of the landed interest without threatening the alkali trade. The Select Committee on Noxious Vapours (1862: x) recognized this when making their report:

> ... The majority of the trade is willing to concur in the object proposed by Lord Derby in his speech, namely, the compulsory condensation of Muriatic gas (Hydrochloric acid), as will enable a measure to be framed which, while protecting the public, will not be injurious to a manufacture occupying so large an amount of capital and labour, so important to the prospects of the country at large, and essential to the actual existence of large communities.

The Alkali Act of 1863 gave effect to recommendations of the Select Committee report. It required that 95 per cent of the hydrochloric acid should be removed from the waste gases. It also established an inspectorate to ensure that the law was enforced and that industry was given advice on pollution control. In the first instance, inspectors had to climb to the top of the chimney to take a sample! The Act achieved a successful outcome for both the landed interest and industry. By 1865 the average escape of hydrochloric acid fumes was 1.28 per cent, and five years later was 0.62 per cent of the amount prior to control. Manufacturers also prospered because new and profitable uses were found for the condensed acid (McLeod 1965).

Production from alkali works doubled between 1862 and 1876 and the production of other chemicals released noxious fumes which were outside the jurisdiction of the Act. Consequently pressure continued for extending legislation. In 1874 the Alkali Act was amended to require that the 'best practicable means' be used to control the fumes from a number of industries, and the hydrochloric acid fumes were also tightened to 0.2 grains/ft^3. Many notable persons had been irritated by noxious vapours, including Queen Victoria herself who had complained that ammonia from a cement works near Osborne was making her royal estate uninhabitable. The Archbishop of Canterbury complained about the deterioration of Lambeth Palace. 'There is not a figure that has got a nose on it' declared the rector of Lambeth, 'or if it has, at a touch it will drop off.'

After 1870 there was a surge of pressure for further legislation. The Lancashire and Cheshire Association for Controlling the Escape of Noxious Vapours and Fluids from Manufacture petitioned the House of Lords, and in 1877 a Royal Commission on Noxious Vapours was appointed. The main problems facing the reformers were the lack of proof that noxious vapours caused ill health and the lack of practicable technology for control. Nevertheless, legislation in 1881 was able to extend inspectorates' powers to cover two more fixed standards for SO_4 and NO_3. It also extended by a dozen or more the list of processes to be registered. For cement works there was no practicable method of controlling the dust but, such was the extent of complaints that the Act empowered the Local Government Board to extend jurisdiction over cement works once an effective measure had been discovered. Legislation between 1863 and 1881 laid the foundations for controlling noxious forms of air pollution. It benefited landlords and farmers by helping to reduce pollution, but it also benefited industry as it legitimated low levels of acidic fumes which could have been the cause of Common Law action. Legislation therefore protected the alkali and other manufacturers from closures such as that forced on Muspratt prior to statutory regulation.

Economic interests in pollution control decision-making in the post-war period

Despite increased participation and 'democratic' procedures in policy-making since the nineteenth century, the pluralist perspective fails to recognize the power of industrial and commercial interests in defining issues and influencing policy. There is little doubt that as government has intervened and extended its field of administration, so too has it been penetrated by corporate interests. For example, while federal government in the US has become increasingly dependent upon advisory committees for advice and legitimation of policy, many of the more important committees are dominated by corporate interests. Steck (1975: 251) described how such committees can have a profound influence on pollution control:

> .... The successful blocking of the industrial wastes inventory for seven years by the Advisory Council on Federal Reports (now the Business Advisory Council on Federal Reports) is only one of the more notorious

examples of the success of advisory committees in frustrating or redirecting information, gathering efforts of pollution control agencies, distorting contracted research studies, watering-down criteria reports or otherwise influencing anti-pollution efforts in hitherto closed advisory sessions.

Typical of the governmental-industrial partnership in the US was the establishment of the National Industrial Pollution Control Council (NIPCC). This committee functioned for four critical years from 1970 to 1973 when there was considerable public support for pollution-control policies. It was composed of top executives from major American corporations and trade associations. It was set up to ensure that industry was properly consulted on pollution control policy at the highest level of government. Steck claims that the NIPCC did not monopolize the flow of information and advice to government, but it did have a privileged position of access which was not available for environmentalists. Hence corporate interests were well positioned to define alternative strategies at an early and crucial stage in the policy-making process.

Despite the superficial appearance of democratic processes of decision-making, it can be claimed that economic forces and interest groups are sufficiently important to influence public opinion, law, and social institutions to favour the dominant classes or interests at the expense of subordinate classes or interests. Pollution control law may therefore reflect sectional interests rather than a pluralistic attempt to promote greater efficiency for society as a whole. The pluralists' emphasis upon consensus, public opinion and behavioural analysis independent of economic interests, while offering only a weak explanation, is a convenient means of legitimising the underlying social order. The question raised and unsuccessfully answered by this interpretation is how and why public opinion on pollution issues rises and falls.

In direct contrast to the pluralist perspective on the formation of issues, Bacharach and Baratz (1963) have suggested that issues are confined to 'safe' areas of public debate. The pluralist view of competition between the most worthy issues is rejected, and instead it has been suggested that public issues are linked to those that will not undermine basic economic and political interests. The pluralist analysis focuses upon issues and the behaviour of participants, with explanatory accounts typically describing the participants involved and the various observable steps taken to tackle the social problem, such as initiating, decision-making and vetoing. Critics claim that while power and influence may be exercised in a limited form in situations of observable conflict, the political agenda may also be controlled without any observable behaviour. Interest groups with little power and influence will tend not to raise issues when expectations of successfully resolving them in their favour are small. The power of subordinate groups may be further constrained by law and institutions. Hence ruling classes may enjoy their power because of passive acceptance of the status quo.

Westergard and Resler (1976) claim that a behaviourist approach to social problems would neglect the domination of a paternalist employer over a passive and unorganized labour force or of a fascist government which had eliminated all the effective opposition. The behaviourist analysis becomes most relevant when the social order becomes disturbed. Molotch and Lester (1973) have shown how accidents can reveal underlying features of the social order which would not be revealed by routine events. Routine events, they argue, are presented to the public in a form compatible with a pluralist mode of society. Accidents and scandals, on the other hand, demonstrate the inadequacy of the pluralist perspective. This is illustrated by reference to the Santa Barbara oil spill (Molotch and Lester 1973: 500):

> ... We gain from the Santa Barbara oil spill a rare view of the oil companies' marriage to the federal government and the effects of that marriage upon local communities. We see how the latter come to be dominated by private decision-making in corporation board rooms and in the office of the Department of the Interior. As upper middle and upper class Santa Barbarans struggle to be heard, to gain access to key decision-makers, they gained even more direct information about power in America and about the inefficiency of local protest. The discrepancy between pronouncement and practice on the part of corporate and federal officials was poignantly illustrated.

In many cases the extent of pollution control policy, especially when there is only weak opposition to dominant capital-forming interests responsible for the pollution, may not be explained simply by reference to behavioural analysis. The power of corporations to prevent pollution control without the issue reaching the political agenda has been clearly demonstrated by Crenson's study, *The Un-Politics of Air Pollution*. His research was aimed at the question: Why had many American cities been slow to develop air pollution control as a political issue? In particular he compared east Chicago, which began to take action in 1949, and Gary which took no action until 1962.

The thesis is that in Gary industrial sources of power came principally from the large US steel company. This concentration of power, he argued, was implicitly recognized by local politicians and consequently no action was needed by US steel to suppress the pollution issue. In Crenson's words: 'Where industrial corporations are thought to be powerful in the matter of anti-pollution policy either the emergence or growth of the dirty air issue is likely to be hindered' (1972: 1222). In east Chicago, on the other hand, industry was more disparate and had less reputation for influence. Hence the air pollution problem was raised and dealt with at an earlier time. When Gary did eventually raise the air pollution issue, largely as a threat of federal state action, Crenson argued that 'US steel ... influenced the content of the pollution ordinance without taking any action on it, and thus defied the pluralist dictum that political power belongs to political actors' (1971: 69–70).

Crenson's thesis rests on the idea that development of issues is dependent upon power reputations of different organizations. Anticipated reactions of others are sufficient to prevent an issue from arising just as anticipated reactions act as constraints on immoral or deviant actions. Here Crenson's notion of reputation for power is closely aligned with the sociological tradition of symbolic-interactionism whereby individuals and groups react to definitions and constraints imposed by others. However, Crenson deals specifically with non-behaviour, and argues that the definition of what is possible and what is not is closely associated with concentration of capital. He supports his argument not by observing behaviour of participants, but by empirical studies aimed at elucidating political leaders' perceptions of industry's reputation for power. From interviews covering fifty-one cities his conclusions confirm his thesis, namely that 'the air pollution issue tends not to flourish in cities where industry enjoys a reputation for power' (1971: 145). Crenson's study is supported by others who have shown that empirical data on pollution abatement reveal that the larger business sectors have been less affected by pollution controls. This suggests that the more concentrated businesses are better able to resist controls that are not in their own interests (Roos and Roos 1972). Walter and Storper (1978) have also shown how the Clean Air Act 1970 has been eroded by relaxation of standards, non-compliance, and relocation to avoid areas with strict regulations. Industries critical to the national economy including the motor car, steel and electric utility industries were in the forefront of opposition. The weakening of air pollution regulations by the economic and political power of industrial capital was enhanced by adverse economic conditions such as the recession and energy crisis.

The same argument that Crenson brings to bear on the development of pollution control policy can also be applied to its enforcement. Reference has already been made to the weak prosecution policy of the Alkali Inspectorate and other enforcement bodies (Ch. 6.) Carson (1971) has shown in some detail how the Factory Inspectorate has also been reluctant to adopt a strong prosecution policy. He looked at the Inspectorate's files relating to a randomly selected sample of 200 firms in one district of southeast England for the period from mid-1961 to the beginning of 1966. Some 3,800 offences were reported, with every firm contributing at least some violations. The minimum figure for violations per firm was two and the maximum was ninety-four. He isolated six types of enforcement decision taken by the Inspectorate on 683 out of the 3,800 recorded offences. No formal action was taken on 5.5 per cent; notification of matters requiring attention was taken on 74.5 per cent; notification of matters urgently requiring attention was taken on 11.9 per cent; indirect threat of prosecution was taken on 4.5 per cent; direct threat of prosecution was taken on 1.8 per cent; and prosecution was authorized on 1.5 per cent.

Pollution control legislation, enforcement and routine decision-making usually help to resolve conflicts of interest. Polluting industry is more likely to have to control its pollution if interests affected are powerful, and the economic consequences serious. Less-well organized interests are not so likely to secure preventive measures. This is well illustrated in Hall's (1976) study of the Port Tennant's Anti-Pollution Association's campaign against the pollution from a United carbon black factory. The campaign was a rare example of a working class and largely female protest. The community action via the Association formed in 1970 was, according to Hall, a result of the strong community structure in the Port Tenant area, the particularly annoying form of pollution which affected furnishings and laundry through the deposition of a fine black oily dust, the twenty years of frustrated attempts to effect pollution control through petitions to the guildhalls and the local MP, and finally the more widespread publicity of pollution problems in general in 1970. The Association blockaded the factory and obtained a good deal of press and media support, but was unable to bring about any major concessions from the company.

In most instances the main opposition to pollution has come not from the working class, but from professional bodies and the middle class. Indeed, frequently, organized labour, management and shareholders all threatened by costs of pollution control have united in opposition. One example of this occurred when the Alkali Inspectorate, local industry and unions opposed the use of World Health Organization air quality standards in the Cheshire County structure plan. The argument of the unions, in this case the Transport and General Workers' Union, has generally been that improvements in pollution control would adversely affect local employment by preventing the expansion of existing companies (Wood 1978). In this and other situations discussed the major influence on decision-making is the maintenance of profit and the requirements of capital within the market economy. In the words of Gunningham (1974: 84–5): 'Any compromise solution to conflict is always resolved, not from within the full range of alternatives which represent the interests of the contending groups, but within a narrow span which favours the interest of capital.'

Conclusions

In this chapter we have looked at the assumptions that lie behind the pluralist perspective on decision-making. This perspective emphasizes the role of public opinion, and many diverse political interests in the development of policy. However, the unequal distribution of influence and power are not consistent with the notion that issues and their resolution are the outcome of public opinion alone. The theoretical weakness of the pluralist perspective is illustrated by two historical examples dealing with the origins of clean-air legislation and early industrial air pollution control in Britain.

Conclusions

The importance of economic interests as opposed to the behaviour of a wider plurality of interests was further demonstrated in more recent issues. Studies that confine their attention to political actors are shown to be inadequate in coming to an understanding of pollution issues. In the next chapter we shall see in much greater detail how the influence of industrial capital has affected the understanding of pollution, the setting of standards and their enforcement.

CHAPTER 8

Pollution in the asbestos industry

Introduction

Asbestos, a name derived from a Greek word meaning 'incombustible' is a generic term for a variety of mineral silicate fibres. There are two groups – the serpentine group containing chrysotile ('white asbestos'), and the amphibole group containing crocidolite ('blue asbestos'), amosite ('brown asbestos'), anthophyllite, as well as actinolite and tremolite which are brittle, have less tensile strength and are little used by industry. In 1978 over six million tonnes of asbestos were produced, of which about 95 per cent was in the chrysotile form. The Soviet Union and Canada are the main producers of asbestos. (Tables 8.1 and 8.2). Chrysotile is mined mainly in Canada, the Soviet Union and Zimbabwe. Amosite occurs only in South Africa. Crocidolite is found in South Africa, Western Australia and Bolivia.

In 1975 the total imports of asbestos to the UK amounted to nearly 140,000 tonnes (Table 8.3) and in 1979 about 25,000 people were estimated to work with asbestos (Advisory Committee on Asbestos 1979a: 19). Asbestos manufacture in the UK is dominated by three firms – Turner and Newall and Cape Industries, which own mines in South Africa, Canada and Zimbabwe, and Belting Asbestos.

Table 8.1 Trends in world production of asbestos

Year	World production (million kg)	% Canada	% USSR
1960	2,210	45	29
1970	3,490	44	30
1973	4,093	41	31
1974	4,115	40	33
1975	4,560	23	48
1976	5,178	29	44

Source: International Agency for Research on Cancer (1977: 29).

Table 8.2 Estimated world production of asbestos by country and region, 1979

	Tonnes
USSR	2,470,000
Canada	1,492,719
Zimbabwe	250,000
China	250,000
South Africa	249,187
Italy	130,000
Brazil	120,000
United States	93,354
Australia	70,000
Others	144,050
	5,269,310

In the US the principal manufacturer is Johns-Manville, whose sales have grown from $40 m. in 1925 to $685 m. in 1971 putting it among the top hundred largest US corporations. By the 1970s in the US 50,000 people worked in asbestos manufacturing alone, and another 40,000 were employed in asbestos insulation work in the building trade (Kotelchuck 1975: 9). It has been estimated that in the US up to a million men and women have worked in the manufacture of asbestos or were so employed in the past. The US Department of Health, Education and Welfare estimated in 1978 that for the remainder of this century and part of next, 17 per cent of all US cancer deaths (over 50,000 per annum) would be attributable to asbestos (Castleman 1979).

History of manufacture and use

Chrysotile was first mined in Canada in 1878. One year later it was being used by Turner to produce yarns in Rochdale, Lancashire. Crocidolite began to be mined in the 1890s, while amosite and anthophyllite were not mined until the First World War. It was not until 1926 that amosite was mined in any great quantity. By then it was being promoted by Cape Asbestos which had already made a success out of its Cape crocidolite mines. In 1890 the world production of asbestos was only about 1,000 tonnes, so it is a relatively new commodity. Its value as an industrial material derives from its strengthening characteristics when combined with other materials such as cement, its insulation and fire-proofing properties, and its resistance to acids. Chrysotile is both strong and flexible and is used in the manufacture of textile products and as cloth for fire-proof protective clothing, fire-curtains; for brake linings, clutch facings and gaskets; and for asbestos-cement products. Amosite is less flexible but is more resistant to heat and acids, and has a higher tensile strength. These properties have made it

Table 8.3 Imports of asbestos to the UK (metric tonnes)

Year	Total asbestos	Chrysotile tonnes	% of total	Crocidolite tonnes	% of total	Amosite tonnes	% of total	Antophyllite tonnes	% of total
1946	54,400	50,700	93	1,000	1.8	2,700	5.0		0
1955	142,100	123,000	87	6,800	4.8	12,300	8.7		0
1965	173,100	147,000	85	3,400	2.0	22,600	13.1	100	0
1975	139,400	120,000	86		0	19,200	13.8	200	0

Source: Figures from Advisory Committee on Asbestos (1979b: 14).

useful for thermal insulation for fire protection and sound insulation. Crocidolite has good filtering and drainage properties as well as mechanical strength, and is still used in certain countries mainly in the manufacture of asbestos-cement pressure pipes. Before 1970 it was widely used in textile products which required acid-resistant properties. Unlike amosite it could be spun, and sprayed for thermal and acoustic insulation. Asbestos has wide applications in manufacturing for laggings, heat and sound insulation, as a filler for products, as a friction material, and as a reinforcer for cement products, washers and gaskets.

The first large-scale use of asbestos occurred at the beginning of the century when commercial techniques were found for spinning and weaving asbestos into a material suitable for insulation. World production grew slowly to 200,000 tonnes in 1920, but began to increase rapidly following the First World War, when asbestos was found to be a useful insulating material in naval vessels. In 1931 a technique for spraying asbestos was developed in Britain. After the war there was a rapid increase in demand for asbestos for clutch and brake linings in the motor car industry, and for construction material in the building industry. Since then the number of uses for asbestos has increased enormously. It is used in some 3,000 products including car battery cases, paper products, roofing material and cement pipes.

In Britain, as elsewhere, the main demand for asbestos is for asbestos-cement products in the construction industry. About 33 per cent of UK asbestos fibre was used for this purpose in 1976. Asbestos is also used in significant quantities for fillers and reinforcements (22 per cent), floor tiles and coverings (12 per cent), and friction materials, especially for motor

Table 8.4 Estimated annual sales turnover for asbestos industry

Product group	U.K. (£ m.)	Export (£ m.)	Total (£ m.)
Asbestos cement	75	5	80
Asbestos insulating board*	2.5	1	3.5
Friction materials	75	30	105
Gaskets, joints and packing	12.5	12.5	25
Textiles, millboard, paper	17	11	28
Flooring	25	24	49
Others – fillers, paints and adhesives etc.	16	4	20
	223	87.5	310.5

Prices at 1 March 1980

* During 1979 asbestos insulating board sales declined steadily in favour of the non-asbestos equivalent and this trend was expected to continue in 1980.

Source: Asbestos Information Centre.

Table 8.5 Asbestos distribution by end use, grade and type in the US, 1974 (million kg)

	Chrysotile	Crocidolite	Amosite	Anthophyllite	Total
Asbestos cement pipe	168	33	0.9	0.18	202
Asbestos cement sheet	82		3.9		86
Flooring products	139				139
Roofing products	66		1.5		67
Packing and gaskets	26	0.09			26
Insulation, thermal	6.6		1.6		8
Insulation, electrical	4.2				4
Friction products	72			0.18	72
Coatings and compounds	34				34
Plastics	15	0.18		0.63	16
Textiles	18				18
Paper	57	0.18			57
Other	33	0.36	0.45		34
Total					763

Source: Clifton (1974).

vehicles (12 per cent). The annual sales turnover for each product group in Britain is shown in Table 8.4. Different demand for asbestos products in the US is shown in Table 8.5. While most asbestos products are made from chrysotile, amosite is used in large quantities for fire-resistant boards. About 84 per cent of amosite is used for this purpose in the UK. However, in 1979 British manufacturers began to favour non-asbestos substitutes when it became clear that amosite dust control standards were to be considerably tightened. Crocidolite was at one time used for a wide variety of products in the UK, but since the 1969 Asbestos Regulations set a 0.2 fibres/ml standard for crocidolite compared with a 2 fibres/ml standard for other forms of asbestos, the fibre has not been used in UK manufacture. Despite this, world production of crocidolite increased to 164,727 tonnes in 1975.

Asbestos dust conditions from mining to disposal of waste

Asbestos becomes harmful only when fibres are released into the air and are inhaled. The use of wet methods of production, when correctly applied, can greatly reduce levels of dust; exhaust ventilation and filtration can help to control dust levels; but when this is impracticable or insufficient to control dust levels people must be provided with protective equipment. Risks from asbestos occur at several stages from mining to disposal of waste products. These include mining itself, transportation, manufacture, handling and use of asbestos products, and finally removal, dismantling, stripping, demolition and disposal.

Little is known about asbestos dust conditions in asbestos mines although they are known to vary from 1 to 9 fibres/ml in some South African mines in 1976 (BSSRS 1979: 63). There are dangers associated with transport of asbestos fibres especially when packed in jute and paper bags as was common practice especially in shipments from the Soviet Union, Italy and India up until recent years. The danger arises from damage to bags and consequent release of fibres (Commission of the European Communities 1977a). In asbestos manufacturing the first stage where risks occur is in opening bags containing the raw fibre and tipping them into a receptacle. When fitted properly with exhaust ventilation equipment dust concentrations should be below 2 fibres/ml. Higher figures were common in the 1960s, but recent installations of suitable exhaust ventilation have reduced this source of dust. Most asbestos cement manufacture is done under wet conditions and dust levels are low. Partial enclosures may also be used to control dust produced in fibre-grinding and milling. In manufacturing processes where it is not possible to use wet conditions or enclosures local exhaust ventilation should be used when possible. It is much less easy to control dust levels in the manufacture of asbestos textiles than in asbestos cement. The dust concentrations sometimes arise above 2 fibres/ml and respiratory protective equipment should be worn. Asbestos dust does arise from sawing, drilling and grinding, but can be controlled by exhaust ventilation and respiratory protective equipment. There are dangers in fitting asbestos cement products and insulation board in building and shipbuilding work, when a considerable amount of hand and mechanical sawing, drilling and shaping is involved, unless there is adequate exhaust ventilation and in some instances use of approved protective clothing and respiratory protection. Machine sawing without effective exhaust ventilation can lead to asbestos dust concentrations of up to 20 fibres/ml (Table 8.6).

There is a risk to the do-it-yourself enthusiast who works with a lot of asbestos but some protection is possible by dampening the asbestos, using hand tools and working in well ventilated spaces. There is also a risk to the

Table 8.6 Concentrations of asbestos dust measured during operations on asbestos cement

Operation	fibres/ml
Machine drilling	less than 2
Handsawing	2–4
Machine sawing without effective local exhaust ventilation	
1. Jig saw	2–10
2. Circular saw	10–20
Machine sawing with effective exhaust ventilation	less than 2

Source: Advisory Committee on Asbestos (1979a: 21).

general public in buildings where asbestos products have been damaged, so releasing asbestos dust into the air. In such cases it may be necessary to remove the asbestos. In some instances asbestos has been removed because of possible future danger. In January 1980 The *Daily Mirror* reported that asbestos was to be removed from the Royal Yacht *Britannia* and any other Navy ships where it was found, because of the assumed hazard. In 1979 in Coventry cladding around heating boxes was removed from 255 council homes in a fifteen-year-old estate at a cost of £500,000 because it contained crocidolite. The operation was undertaken despite the lack of any evidence of risk at the time to the residents. It was carried out under almost siege conditions – all the residents were evacuated to caravans during the three-month operation (*Safety* Jan. 1980).

In Britain one of the greatest problems of dust control today occurs in dismantling and demolition work where it is less easy to apply dust-control measures, and where the non-static nature of the work and the numerous firms involved make supervision of work particularly difficult. A preliminary study indicates that about 40 per cent of those who have received compensation due to asbestosis in the period 1961–75 were involved in lagging and delagging (Health and Safety Executive 1978). Finally there is a risk from waste disposal. Waste asbestos should be bagged and disposed of in such a way as to prevent rupture of the contents, and the bags should be covered by soil or non-toxic material within twenty-four hours.

One of the most worrying global trends is the growth of asbestos manufacture in countries with non-existent or weak controls over occupational health risks. This has been influenced by the widespread public recognition of the threat of asbestos in industrially developed countries. Mexico's, Taiwan's and Brazil's exports of asbestos-textile products to the US rose from 82 kg in 1969 to 1.3 million kg in 1973 (Kotelchuck 1975: 14). Large multinational companies may decide to carry out the most hazardous processes in countries with the weakest controls. During the 1970s the British-based Cape Industries no longer used blue asbestos for manufacture in Britain but continued to mine it in greater quantities in South Africa (BSSRS 1979: 63). This aspect of asbestos pollution requires an international solution, the character of which was discussed in Chapter 4.

The hazards of asbestos revealed – an historical perspective

No records exist of what happened to the South African workers who were first involved in the mining of asbestos on a large scale in the 1880s. The risks of manufacture must have been apparent in European factories by the turn of the century. Of seventeen people first employed in a French factory, only one had survived five years of work. Auribault (1906) observed that there had been numerous deaths among French asbestos spinning and weaving workers, but was unable to suggest a reason for the excessive number of deaths. The first recognized death due to asbestos was diagnosed by a London physician in 1900. The case study was based on an autopsy of a thirty-three-year-old man who had been working in a carding room of an asbestos textile mill. The lungs were severely scarred and contained numerous asbestos fibres. The case was later reported in 1907 by the Departmental Committee on Compensation for Industrial Diseases.

The medical profession was slow to investigate and the issue remained dormant until the mid-1920s. The business world, however, was not so slow to recognize the dangers of asbestos. In 1918 asbestos workers were refused personal life insurance policies by US and Canadian insurance companies (Kotelchuck 1975). Lack of medical interest can be explained by the long latency period between first contact with asbestos and symptoms of asbestos-related disease, and the low levels of asbestos in use before the First World War. With growth in the asbestos industry, workers' ill health and death became more commonplace. The investigation of asbestos-induced disease was again reported in 1924 when Cooke described in the *British Medical Journal* a case of fibrosis of the lung found in the autopsy of a thirty-three-year-old woman who had worked in an asbestos factory. In 1927 Cooke described the medical condition as one of asbestosis. By the end of the 1920s twelve such cases had been reported in Britain and in 1930 the first case was reported in the US. Five years later there had been twenty-eight asbestosis cases reported in the UK and the US.

Symptoms of asbestosis were also reported. Earliest clinical symptoms were a slight persistent dry cough, shortness of breath under mild exertion, finger clubbing, lack of energy, frequent chest infections and loss of weight. As the lung tissues become more scarred by the asbestos fibres insufficient oxygen enters the blood, leading to more and more severe breathing difficulties and in the end suffocation. Epidemiological studies demonstrated the widespread nature of the disease. Merewether and Price conducted a wide-ranging survey of asbestos-workers in England, and their findings were reported in 1930. This report, known as the Merewether Report, found that 80 per cent of workers who had been in contact with asbestos twenty years or more previously had symptoms of asbestosis. A clear correlation was demonstrated between the duration of exposure and concentration of dust to the time taken for the disease to develop. The longer the period of exposure and the higher the concentration of dust the shorter the period before the development of asbestosis symptoms. Similar findings in Germany were later reported in 1959 by Böhme. His study showed that rates of asbestosis symptoms increased with the length of exposure, reaching a peak rate of 79 per cent for workers exposed for more than ten years.

In 1935 a possible link between asbestos and lung cancer was reported by two doctors from South Carolina. Other autopsy cases confirmed the association of asbestosis and lung cancer. However, according to Kotelchuck (1975) the asbestos industry successfully defused the health concern by funding research projects likely to give a favourable picture of the asbestos industry. In 1938 two scientists from Saranac, Vorwald and Carr, argued that the association between asbestosis and cancer merely demonstrated that those who suffered from asbestosis were more susceptible to lung cancer. This was an argument which could not be countered using autopsy data alone.

During the twenty-five years between 1935 and 1960, which witnessed a massive growth in asbestos use in the world economy, the issue of health risks smouldered away. By 1960 there had been a total of sixty-three papers in the US, Canada and Britain on asbestos-related disease. Eleven of these papers that were sponsored by industry denied that asbestos was a cause of lung cancer and played down the seriousness of asbestosis. Typically, industry-sponsored reports were biased, carefully including high proportions of relatively young groups of workers. Kotelchuck suggested that industry-sponsored epidemiological research tended consequently to hold back knowledge about the harmful effects of asbestos dust. However, a study by Turner Brothers' company doctor in 1955, reported in the *British Journal of Industrial Medicine*, did show that British asbestos workers were ten times more likely than the average worker to get lung cancer (BSSRS 1979: 25).

The fifty-two papers that had not been sponsored by industry painted a rather different picture. Based in the main on autopsies, these papers suggested that asbestos might be the cause of lung cancer (Kotelchuck 1975). One such report in 1947 was based on an epidemiological study undertaken by Merewether, then Chief Inspector of Factories in Britain. The survey was of 235 people known to have died of asbestosis between 1924 and 1947. Thirteen per cent of the victims were also found to have had lung cancer. This compared with a much lower cancer rate of 1.32 per cent for victims of silicosis during the same period. The association between asbestosis and lung cancer was confirmed later by a similar study reported in 1951 which looked at 1,205 cases of pneumoconiosis (lung diseases caused by dust). The report revealed that 14 per cent of those who had suffered from asbestosis also had lung cancer. In the case of silicosis only 6.9 per cent had lung cancer (Ahmed *et al*. 1972).

In the early 1960s, after a massive growth in asbestos production and use, there was a much clearer picture of asbestos hazards. In 1960 a new and invariably fatal disease, mesothelioma, was recognized by Wagner and his associates to be causally linked with asbestos. Mesothelioma, involving malignant tumors of the lung and abdominal cavity lining, causes acute pain and breathlessness. Most people die within two years of first diagnosis. Eighty-five per cent of cases have been directly attributed to asbestos exposure, the only known environmental cause of the disease. There appears to be no safe limit as several cases of the disease have occurred after only brief or low exposure to asbestos dust. In some instances the disease has been contracted by relatives of asbestos-workers and those living near a source of asbestos dust such as in the vicinity of asbestos mines, asbestos-processing factories and asbestos-dumps (Commission of the European Community 1977a). Mesothelioma has been common in people exposed in and around the crocidolite mines in South Africa and Western Australia but is rare in the amosite and chrysotile mines. In a study of mesothelioma cases notified in England, Wales and Scotland in 1967–8, 68 per cent had definite occupational exposure, 7 per cent possibly had occupational exposure, and 5 per cent had neighbourhood – domestic or hobby – exposure. In 15 per cent no past exposure could be shown and in 5 per cent no case history could be obtained (Greenberg and Lloyd Davies 1974). There were 292 death certificates issued mentioning mesothelioma in Britain in 1976 (Advisory Committee on Asbestos 1979a). To what extent this represents the complete picture is still uncertain. Exposure to asbestos has also been shown to increase the rates of other cancers of the stomach, colon, oesophagus, rectum and larynx.

Concern about asbestosis also increased when a survey in 1963 of 500 autopsies on urbanwellers in Cape Town revealed that 26 per cent of the lungs were affected by the disease. Several other autopsy studies have found asbestos fibres in at least 25 per cent of the adult population in urban areas of industrialized countries. In the early 1960s, Selikoff's group at Mt. Sinai Medical Center in New York produced an important survey of asbestos-insulation workers, and from this a clearer understanding of the health risks began to emerge.

Earlier autopsy studies showing an association between asbestosis and cancer had failed to prove definitely that asbestos was a cause of cancer. Selikoff's contribution to knowledge of risks largely derives from epidemiological investigation of a well defined population. Unlike many earlier studies sponsored by industry, investigation was made of workers with twenty or more years since first exposure. The study looked at expected and observed deaths among 632 New York and New Jersey asbestos-workers, who were employed in the industry on 31 December 1942, during the period from 1 January 1943. The study revealed considerably higher death rates than mortality tables would have suggested. Higher incidences of lung cancer, mesothelioma, asbestosis and some other asbestos-related diseases accounted for 421 deaths between 1943 and 1971 as opposed to an expected 280 deaths (Table 8.7).

Another disturbing insight into the hazards of asbestos was the revelation by Selikoff's group that there was a synergistic effect of cigarette smoking and asbestos exposure on the incidence of lung cancer. It was shown that American asbestos-insulation workers who smoked were ninety-two times as likely to die from lung cancer as non-smoking members of the public who had no exposure to asbestos dust. This compares with the

Pollution in the asbestos industry

Table 8.7 Expected and observed deaths among 632 New York and New Jersey asbestos-workers exposed to asbestos dust twenty years or longer (1 January 1943 to 31 December 1971)

	Expected (E)	Observed (O)	O/E × 100
Total deaths:			
all causes	280.0	421	150.4
Total cancer:			
all sites	47.2	189	400.4
Cancer of lung	10.0	84	840
Pleural mesothelioma	*	8	
Peritoneal mesothelioma	*	24	
Cancer of stomach, colon, rectum, oesophagus	13.0	41	315.4
Cancer of all other sites combined	24.1	32	132.8
Asbestosis	*	33	
All other causes	232.8	199	85.5

* Rare causes of death in the general population

Source: Figures adapted from Ahmed et al. (1972: 18).

Turner Brothers' study in 1955 which showed that British asbestos-workers were ten times more likely than average to suffer from lung cancer (BSSRS 1979: 25).

Studies in the 1960s and 1970s confirmed without doubt the serious hazards of asbestos dust. They confirmed the long latency period, frequently between twenty and forty years from initial exposure to symptoms of malignancy. Other studies also demonstrated the public health hazard to those not occupied in the asbestos industry but nonetheless exposed to low levels of asbestos dust. The main risk was of contracting mesothelioma. Recently mesothelioma lung damage was discovered in inhabitants of Turkish villages near naturally occurring zeolite and chrysotile deposits. Damage occurred even though no industrial processes were involved (Martin 1980).

In recent years some of the controversy has shifted to the relative health risks of different fibre types. The final report of the Advisory Committee on Asbestos (1979a) claimed that exposure to crocidolite has been more dangerous than the exposure to chrysotile. Risks from amosite were considered to lie between those of chrysotile and crocidolite. This view, now widely held in Britain, is based on the claim that mesothelioma is mainly associated with crocidolite. Cape Industries' official view, as expressed to the Advisory Committee on Asbestos, was that the dangers of crocidolite in comparison to other fibres could not be justified on scientific grounds. Certainly the distinction between risks is not officially recognized in many countries including West Germany, East Germany, Italy, France, US, the Soviet Union, Canada, South Africa and Norway (Table 8.8). The Advisory Committee admitted the evidence is inconclusive. Critics (Singer 1979) argue that given this uncertainty it is not surprising that crocidolite is blamed as it can be more easily substituted and chrysotile is the most widely used asbestos fibre, accounting for more than 90 per cent of the world's production. Perhaps the official view of Cape Industries that crocidolite is not so dangerous as has been made out cannot be divorced from their crocidolite mining interests in South Africa. Certainly the fact that the use of crocidolite in the UK was already declining in the 1960s and that substitutes can be found without too much difficulty (chrysotile has been used instead but is not so resistant to acids) made it a useful scapegoat when uncertainty prevailed.

Influence of industrial capital on knowledge and understanding of risks

Not surprisingly a good deal of effort is made by the asbestos industry to reduce alarm. Funds have been used to support medical hypotheses which put the blame on other than the chief manufacturing processes and the use of products. The Asbestosis Research Council was founded by British asbestos manufacturers in 1957 to undertake research and to give advice and reassurance to the general public and those in the building and construction industries. While only eleven industry-sponsored papers were published before 1960, the next eighteen years saw nearly seventy in the UK alone. The asbestos industry often claims that present-day risks are minimal as current cases of disease are related to heavy exposure in the past. Improvements in dust control have, it is claimed, reduced these risks. Chissick (1979: 127–8) supported this argument:

> ... following the Merewether Report and the pooling of information and experience by the manufacturers, there were rapid and significant improvements in methods of dust control – enclosing machinery, damping the work, extracting dust from the breathing zone of operators and greater use of the industry covered by the Asbestos Regulations produced a significant reduction in the incidence of asbestos disease in the factories.

This claim is an extravagant one. According to a mortality study by Howard *et al.* (1976), workers in an English asbestos factory who started work after 1950, the period of improved industrial engineering control technology and regulation, had suffered a lung cancer death rate 1.9 times that expected after fifteen or more years from initial exposure. This did not represent an improvement on those who had been in work from 1933 to 1950.

Another argument by one section of the industry is that the manufacture and use of asbestos-cement products is safer than that of other products for four reasons: the asbestos content is only 10 per cent; the fibre used is

Table 8.8 Present hygiene standards for asbestos in fifteen countries

Present hygiene standards for asbestos

Country	Type of asbestos	If known TWA in hours*	No. of fibres/ml or equivalent (unless otherwise stated)	Legislation or guidelines	When legislation took effect (if known)
Australia	All		5 million parts ft³	Harmful gases, fumes, mists, smoke and dust regulations	18.4.1977
Belgium	All		2	Law in preparation	
Canada	All	8 hrs	5 Quebec	Each state has set of rules e.g. Quebec has laws and regulations covering health and safety in mines and quarries	
		8 hrs	2 Ontario		
Denmark	All		2	Law	1972
France	All	Working day	2 1	Decree	20.10.77
Germany (Federal Republic)	Dust 100% chrysotile Dust less than 3½% chrysotile (NB Precise limit in a given workplace depends on how much chrysotile there is in the dust when analysed by weight)	Breathing zone Air of workroom	0.15 mg/m³ 4 mg/m³	Interlocking requirements from federal government, Landergovt. Central Union of Trade Co-operatives	
Ireland (Republic of)	Crocidolite Other		0.2 2	Regulations	1972
Italy	All		5	Guideline only	

136

Influence of industrial capital on knowledge and understanding of risks

Country	Type	TWA*	Limit	Legislation	Date
Netherlands	Crocidolite		Less than 2	Decree	1.4.78
	Others		2		
Norway	All		5	Regulations	1973
Sweden	All (crocidolite banned)		1	Regulations	1.7.76
South Africa	All			Occupational diseases in Mines and Works Act	1973
USA	All	8 hrs	2		
	Ceiling for all employees		10		
USSR	All Dust<10% asbestos		2 mg/m³	Regulations	1.7.76
	Dust<30% asbestos		1 mg/m³	Regulations	
UK	Crocidolite	10 mins	0.2	Regulations supplemented by guidance notes	1970
	Other	10 mins	12		
		4 hrs	2		

*TWA = Time weighted average.

Source: Advisory Committee on Asbestos (1979b: 95).

mostly white asbestos; the manufacturing process is almost entirely wet and hence dust levels are low; and the asbestos fibres are encapsulated in cement. Even if some of these points are valid it should not detract from the fact that in one asbestos-cement plant alone in the US seventy-two cases of mesothelioma had been observed up to 1973 (BSSRS 1979: 46). In Sweden recognition of risks led to a ban from July 1976 on asbestos-cement products with certain exemptions for asbestos-cement pipes.

The extent of mortality from asbestos-related diseases is probably substantially underestimated by records from death certificates. Lung cancer victims who smoked may not have asbestos recorded on the death certificate. According to British death statistics mesothelioma deaths from asbestos are more common than those from lung cancer (Table 8.9). However, studies of excess mortalities found in industrial populations suggest that lung cancer is a more common cause of death from asbestos than is mesothelioma (Table 8.10). Official cases of asbestosis as diagnosed by the Pneumoconiosis Boards indicate a growing incidence of the disease. The number of cases reported each year has grown from 134 in 1969 to 189 in 1976. The rise in reported cases cannot of course be dissociated from the growth in numbers of workers receiving compensation, the number receiving government compensation rising from 53 in 1961 to 189 in 1976 (BSSRS 1979). These figures no doubt reflect a growing awareness of the possibilities of seeking compensation. The Society for the Prevention of Asbestosis and Industrial Diseases (SPAID) has been influential in this respect. There is now a greater proportion of people rejected by the Pneumoconiosis Boards, and there is a growing disagreement between hospital consultants and chest clinics on the one hand and the Pneumoconiosis

Table 8.9 Death certificates mentioning mesothelioma, etc. UK, 1967–76

Year of death	Mesothelioma	Lung cancer with mention of asbestosis or asbestos exposure	All certificates mentioning asbestosis
1967	88	34	60
1968	153	25	81
1969	159	25	77
1970	193	24	80
1971	176	32	78
1972	208	40	103
1973	218	41	102
1974	221	41	120
1975	256	49	163
1976	292		

Source: Health and Safety Executive (1977).

Table 8.10 Excess of observed minus expected deaths. Based on 8,327 deaths from industrial cohorts reported in eleven studies

	Observed – Expected deaths
All causes	1,407
Cancer of lung	589
G I cancer	121
Mesothelioma	79
Asbestosis	367

Source: Figures are taken from Advisory Committee on Asbestos (1979a: 51).

Boards on the other as to criteria for assessment. The official figures can therefore only represent a small proportion of those affected.

What is clear is that there are no known cures for any of the diseases caused by asbestos, and the solution to the problem is the prevention of exposure to asbestos dust. Having reviewed the development of knowledge about health risks, we can now go on to look at the development of asbestos regulations, at how the balance between risks, costs and benefits has been assessed, and how regulations have been introduced and enforced.

Asbestos regulations in the UK and elsewhere and the dose-response relationship

In the last section it was found that asbestosis was recognized by several doctors in the late 1920s to be associated with exposure to asbestos dust. In particular the Merewether Report not only provided an understanding of the asbestosis hazard, but also made recommendations about how it could be prevented. It recognized that the main task in preventing the disease lay in reducing the concentration of asbestos particles breathed during the manufacture of asbestos. This could be carried out by improved ventilation and dust suppression. Practical problems were discussed by the Factory Department of the Home Office and the asbestos-factory owners. This led to a number of agreements upon which the General Council of the Trades Union Congress were invited to make comments. These discussions led to the Asbestos Industry Regulations 1931, under section 79 of the Factory and Workshop Act 1901

The general requirements of the Regulations were to prevent the escape of all asbestos dust into the air where people worked by the use of exhaust ventilation systems, and to provide breathing apparatus and protective clothing when it was impossible to avoid dust during certain manufacturing operations and machinery-cleaning. Certain processes were exempted from control on the grounds that asbestos dust was not a problem. The regulations were insufficiently comprehensive and were conceived in a way that led to great difficulties in reducing risks from asbestos. They did not apply to all occupations carrying a risk of exposure to it; they did not cover all activities in asbestos manufacture which nevertheless could give rise to health risks; they took no account of mesothelioma and lung cancer hazards, then unknown; they did not attempt to control public health risks that might exist; and they provided an absolute ban on asbestos dust which proved unworkable and unenforceable.

In the US the first standard for asbestos dust was based on a 1938 study of 541 workers in several asbestos-textile mills. The study was one of those that had severe limitations resulting from inadequate recognition of the latency period and exposure time. Only two had worked for more than twenty years, 66 more than ten years and 333 out of 541 had worked in the mills for less than five years. As a consequence the dangers of asbestos dust were severely underestimated. The American Conference of Government Industrial Hygienists (ACGIH) made use of the study to set a Threshold Limit Value (TLV) of five million particles per cubic foot (mppcf) (approximately 30 fibres/ml). Not surprisingly this standard, based as it was on spurious investigation, was totally inadequate. The standard remained operative until 1970 when a 12 fibre standard was adopted following the Federal Occupational Safety and Health Act.

Concern in the UK about asbestos grew in the 1960s as its carcinogenic properties were discovered. Moreover the number of new cases of asbestosis began to increase alarmingly both in parts of the asbestos industry covered by the 1931 Asbestos Regulations and also in those not covered by them. A special Medical Advisory Committee was set up in 1965 to investigate the problem. Its report in 1968 recognized that there was no medically known safe threshold and pressed for the establishment of a provisional standard based on what was obtainable by the most diligent factories.

The British Occupational Hygiene Society (BOHS) also reviewed the dangers of asbestos and in June 1968 recommended that a limit of 2 fibres/ml should be adopted for chrysotile. This standard would not eliminate the risk of contracting asbestosis, it was admitted, but it would reduce the risk to about 1 per cent if exposure were limited to this level over the working life of fifty years for the asbestos-worker. This conclusion was based on a single study of asbestos-textile workers in employment in 1966 at Rochdale. The Rochdale study was sponsored by the asbestos industry and began in 1951 when measurements of dust were made and regular health inspections were carried out. The investigators, Knox and Holmes, selected 290 men who had been working for Turner Brothers for at least ten years, (but only 118 had worked for longer than twenty years). According to the researchers only eight showed evidence of asbestosis. In a group of eighty men who had been exposed to low levels of dust, only one had asbestosis.

In 1969 Asbestos Regulations were introduced which implemented the recommendations of the BOHS report. A 2 fibres/ml standard when measures are averaged over four hours or 12 fibres/ml when measured over ten minutes was set for all asbestos dust types except crocidolite. The standard for crocidolite was 0.2 fibres/ml when measured over ten minutes. The Regulations required that because of the high risk twenty-eight days' notice of any work involving crocidolite be given to the District Inspector of Factories. Exhaust equipment had to be inspected once a week and thoroughly tested by an expert every fourteen months. The 1969 Regulations covered activities which had been excluded from the 1931 Regulations, and also introduced procedures for the control of waste asbestos.

Following the BOHS report and the 1969 Asbestos Regulations similar regulations were adopted in other countries. In 1972 the US National Institute of Occupational Safety and Health (NIOSH) recommended a basic 2 fibres/ml standard for all asbestos fibres, and this was accepted in the same year by the Occupational Safety and Health Administration of the Department of Labour, but was not introduced until 1976. However, the regulations adopted in Britain, the US and elsewhere were based on evidence that has since been much criticized. In 1970 Lewinsohn, an employee of Turner Brothers, re-examined the same 290 men who had been studied by Knox and Holmes four years earlier. He found that nearly 50 per cent of the group had developed significant lung changes which could lead to asbestosis. This classification was known as crepitations and was distinguished from cases of possible asbestosis and certified asbestosis. These results were first disclosed in 1972, but only in the *Journal of the Royal Society of Health* where they caused little stir. Fortunately the study was picked up by Selikoff, and he proceeded to question the British safety standard in public. According to the Advisory Committee on Asbestos, the level of risk in the first Rochdale Studies, which was the basis for the 1969 Regulations, had been underestimated by a factor of about 15. Another limitation of the Asbestos Regulations was that they were based on risks of asbestosis and deliberately ignored the cancer risk. Yet in the US and elsewhere it has been shown that most excess deaths among asbestos workers are due to cancer and not asbestosis.

Concern about mesothelioma, the rising incidence of asbestosis and widespread public outrage at the conditions at Acre Mill near Halifax, run by Cape Asbestos (now Cape Industries) between 1939 and 1970, led to further inquiries. Allegations of maladministration by the Factory Inspectorate, prolonged disrespect for asbestos regulations, and a clandestine suppression of information about the risks connected with asbestos were documented in a *World in Action* programme, 'Killer dust: a standard mistake', in October 1974 and a BBC 2 *Horizon* programme in January 1975. According to reports from workers, extractor motors frequently broke down, carding machines and roving frames had no dust extraction at all, and while people were issued with respirators they were infrequently worn because dust got in round the sides.

Madden, the local MP, successfully pressed for an inquiry by the Ombudsman. His report in 1976 was confined to questions of the Factory Inspectorate's administration. Nevertheless it revealed a long history of neglect and lack of determination by the Factory Inspectorate to put pressure on the management of Acre Mill to reduce the risks. The question of enforcement of the law will be dealt with later, but the immediate response to the report was for the government to allay public concern by appointing an Advisory Committee on Asbestos to review the risks to health from exposure to asbestos and to make recommendations as to whether tighter standards for protection were required.

The response by the asbestos industry was to launch a £500,000 sponsored counter-attack undertaken by the Asbestos Information Committee. This Committee was formed in 1966 in response to a rise in concern about asbestos disease in the docks. In 1977 it was renamed the Asbestos Information Centre. There was some very questionable information in the first series of full-page advertisements in July 1976. Indeed claims made in these were so extravagant that protests, which were subsequently upheld, were made to the Advertising Standards Authority. The purpose of the advertisements was to allay public concern, and in this they certainly succeeded. Prior to the campaign about one-fifth of the UK population thought that asbestos should be banned. Several weeks after the advertisement this figure had been halved (BSSRS, 1979: 60).

In one of the brochures released to the public by the Asbestos Information Centre called *Miracle fibre-killer dust?* the reader was told that risks were satisfactory because the 1969 Asbestos Regulations were based on scientific evidence. It did not point out that the study at Rochdale has since been discredited. It was stated

> There are very good grounds for the industry's confidence in the efficacy of the 1969 Regulations where they are properly applied. Not only were they based on a study of 290 workers over periods ranging from 10 to 30 years or more but a carefully monitored group of 58 workers who have worked for at least 10 years since 1951 in measured conditions that conform to present regulations show not a single case of asbestos related disease.

The risks from asbestos are in this way blatantly played down.

While the asbestos industry poured money into publicity, research expenditure for 1975–6 at the Asbestosis Research Council, sponsored largely by the asbestos industry, was a meagre £87,000. Moreover, it can be argued that even the research arm of the asbestos industry was aimed at whitewashing the potential risks of asbestos, and at playing down the irresponsibility of the asbestos industry. Cross, Chairman of the Environmental Control Committee of the Asbestosis Research Council, in marked contrast to the Ombudsman's report, responded to a student's inquiry in January 1975 as follows:

> It is true that the company at Hebden Bridge was never prosecuted. The reason for this is that no breaches of the 1931 Regulations were, in

fact, observed. Indeed the measures taken for the control of asbestos dust at that factory were, in the light of knowledge available at the time, considered to be very good and the techniques of control applied to many of the operations were, in fact, used as demonstrations of the effective application of ventilation equipment.

Medical arrangements at the factory were, in fact, written up in a booklet published by the Ministry of Labour in 1960 as a good example of well organized medical services in factories.

Although Cape Industries continue to deny that they broke regulations at Acre Mill, other than the vaguely expressed requirement to prevent the escape of dust by ventilation equipmment (Advisory Committee on Asbestos 1977), they have had to pay out considerable sums in compensation. Between 1939 and 1970 2,199 people worked at Acre Mill. According to Cape Industries there have been seventy cases of asbestosis and three cases of mesothelioma among former workers. However, many former workers, like Hadley, suffer from asbestosis but are not certified by the Pneumoconiosis Medical Panel for disability benefit. Pickering, solicitor of the Asbestos Action Group which was set up in 1975, estimated that 400 or more people in all could be affected, including those who had already died. In 1973 Cape Asbestos (Industries) made provision of £3.5 m. relief to cover all future payments. Since then this reserve fund has had to be considerably enlarged (*Safety* Oct. 1979).

It is not surprising that critics of the asbestos industry complain bitterly that most of the information available is controlled by the asbestos industry through the Asbestos Information Centre and the Asbestosis Research Council. According to the BSSRS (1979: 93) 'it has got to such a state that lawyers acting for asbestosis sufferers find it hard to get a doctor with specialist knowledge on asbestos to testify for them'.

The Advisory Committee on Asbestos has made a number of reports on health hazards and precautions. In 1979 it produced its final report which recommended that an appropriate balance between costs and risks would require a tightening of the standard to 1 fibre/ml for chrysotile, 0.5 fibres/ml for amosite, and a total ban on crocidolite. A 0.2 fibre/ml control limit should, it was argued, be applied to work on crocidolite still in circulation. In circumstances when these standards cannot be met, such as in dismantling operations, protective clothing should be worn, suitable changing and washing facilities provided and a ban placed on workers' protective clothing being cleaned at home.

In recognition that there was no absolutely safe level of asbestos dust the Committee preferred to speak of a control limit rather than a standard which might be confused with a safe threshold. The Committee believed that the new 1 fibre/ml standard for chrysotile would still allow between 0.02 and 2.25 per cent excess deaths among workers exposed to this level over a period of fifty years (Advisory Committee on Asbestos 1979b: 76).

Pollution in the asbestos industry

The control limits applied to asbestos fibres which could be observed through the light microscope. The recommended technique uses a membrane filter method which involves drawing a known volume of dust-laden air through a membrane filter and then counting the asbestos fibres collected. Some researchers, however, believe that small fibres not detected by the light microscope could prove dangerous. The Committee also recommended that the Health and Safety Executive should issue general advice on substitutes but not enforce substitution. The recommendations require any person who produces specifications for, or carries on a process, involving the use of asbestos or any product containing it, to consider its substitution by other materials if it is practicable to do so, and if it is significantly less hazardous. Other recommendations included training to a specified standard for those who test ventilation equipment, a streamlining of medical surveys and better controls on asbestos emissions affecting the public.

Reactions to the recommendations of the Advisory Committee on Asbestos were mixed. The manufacturers said they would try to delay the proposal to implement new standards by 1 December 1980. They would also resist pressures placed upon asbestos-users to look for substitutes. The TUC and some environmental groups, on the other hand, had pressed for tighter standards. The TUC had wanted a ten-fold reduction in the standard for chrysotile and a programme of compulsory substitution.

The Asbestos Information Centre, the public relations arm of the industry, said it would contest the control limit on amosite as it believed that it could not consistently be achieved in the manufacture and use of fire-resisting board containing this fibre. Higher standards for amosite would particularly affect Cape Industries, the leading amosite-user in Britain. Although Cape Industries sold their amosite mines in South Africa in 1979 they still have a strong working relationship with the mines importing company in Britain, Cape Asbestos Fibres (*Safety* Oct. 1979). A tighter amosite standard is likely to lead to greater opposition than that imposed on crocidolite in 1969 because its use has been increasing rapidly. Britain imports over a fifth of the world production of amosite. Imports since the Second World War reveal a seven-fold increase for amosite as opposed to a two-fold increase for chrysotile. In 1975 19,200 tonnes were imported which was over three times the largest figure for crocidolite since the war. It was not possible to substitute chrysotile for amosite in insulation board, whereas crocidolite had been more easy to substitute. There are, however, other substitutes and Cape Industries has already decided to stop using amosite for this purpose. Asbestos manufacturers are not so likely to oppose the 1 fibre standard for chrysotile as they have become resigned to a tightening up of this standard since the growth of public alarm in the mid-1970s. Indeed at the time of the final report of the Advisory Committee on Asbestos in 90 per cent of manufacturing and processing plants dust levels were below the 1 fibre limit.

The possibilities of substitution are of great importance as a strategy of pollution control, but it needs to be shown that the substitute can do the same job adequately and is free from significant hazards, also that the costs of substitution are less than costs associated with the manufacture and use of asbestos products. As yet this task has received little serious attention. Calculation of costs involved in substitution are complicated by the fact that no substitute is likely to have exactly the same properties. One of the features of asbestos which has made it an important economic commodity has been its versatility and range of useful properties. In some cases, such as reinforcement of cement, substitutes exist but do not have such a wide range of positive features (Table 8.11). The Asbestos Information Centre has expressed the risk and cost equations in the following terms: 'Many people who would otherwise have perished in fires or on the roads are alive today because of asbestos, yet it cannot be denied that a relatively small number have been made ill or died as a result of past exposure to excessive concentrations of dust.' This is clearly a subjective assertion and there is a general lack of reliable information on the costs and benefits of substitution.

The Advisory Committee on Asbestos (1979b) reviewed the current state of knowledge on substitution. Among their various evaluations it was claimed that asbestos ventilation ducts could be substituted by the use of sheet materials such as plasterboard, metal, plywood, fireboard and plastic, but these can be less convenient and cost more. Asbestos insulation board can be substituted by asbestos-free board such as Supalux, Monolux and Limpet board but the cost would be about 18 per cent more, and there would be greater costs of installation. Asbestos-fibre millboard and paper could be substituted, at great cost for some uses, by aluminium silicate fibre products. Water, supply pipes above eighteen inches in diameter could use iron pipes instead of asbestos-cement pipes, but this would involve greater cost and maintenance. Smaller-size pipes can be substituted by pvc pipes. Asbestos cloth for fire-fighting clothes can be substituted by the use of nylon fibres but again the costs are very much higher. Moreover, available substitutes do not at present combine in one material the heat, flexing and abrasion resistance of asbestos textiles. For a limited number of asbestos products there do seem to be reasonable and competitive substitutes. There are acceptable alternatives to asbestos rope, thermoplastic tiles and pvc (vinyl) asbestos tiles. On the other hand there appear to be no adequate substitutes for asbestos in friction applications in motor vehicles, nor for asbestos/resin laminates used in bearings subject to high temperatures and pressures, chemical attack and water absorption. The case for a programme of enforced substitution is not therefore without its problems, but much more effort is needed to investigate the possibilities and cost of substitution.

In several countries asbestos has been banned from a limited number of uses. In 1972 there was a ban in New York on the use of asbestos in spray

Table 8.11 Fibre reinforcement of cement

	Cost	Fire resistance (ability to hold together up to 1,000°C)	Resistance to alkali attack	Reinforcing characteristics modulus of		
				Tensile	Rupture	Impact
Asbestos	Low	Excellent	Excellent	Excellent	Excellent	Fair
Glass	High	Poor	Variable	Good	Good	Good
Cellulose	Low	None	Poor	Poor	Poor	Good
Mineral wool	Low	Good	Poor	Fair	Fair	Fair
Thermoplastic (nylon/terylene polyethylene)	High	None	Good	Fair	Fair	Good
Carbon	Very high	Excellent	Good	Good	Good	Good
Steel	High	Good	Will corrode in time	Good	Good	Good

Source: Asbestos Information Committee.

fire-proofing on all construction sites. Spraying of asbestos is totally banned in Sweden, the Netherlands and Norway. The main companies in Britain have stopped using spraying techniques, and the Advisory Committee on Asbestos has recommended making it a legal offence (1979b: 99). Asbestos has also been banned in Sweden for cushion flooring, the undersides of mats, paint, glue, putty or jointing material, and in certain asbestos-cement products. Use of crocidolite is banned for all products. In 1979 regulations in Sweden were tougher than in other countries, with a limit of 1 fibre/ml. In Norway by way of contrast, the limit was only 5 fibres/ml. In Canada each province had its own set of rules and regulations. Ontario had a 2 fibre standard averaged over eight hours while Quebec had a 5 fibre standard averaged over the same period of time (Table 8.8). The EEC has also made general recommendations to its member countries. In 1977 the Committee on Environment, Public Health and Consumer Protection reported on the health hazards of asbestos and recommended that the use of crocidolite, and spraying of other types of asbestos should be banned. It also suggested that standards (unspecified) should be based on cancer risks.

It has been implied that asbestos regulations have been set in response to a known dose/response relationship. In reality, however, there is still considerable doubt and uncertainty about this relationship. Given the long-term and cumulative factors in asbestos-related disease it is perhaps not surprising that it is hard to measure the dose part of the equation. The problem is aggravated by the use of different methods of estimating levels of asbestos dust. Before 1970 most sampling was based upon instruments held in a fixed position. Since 1970 samples collected from workers have been used. Differences between the sampling techniques suggest an underestimation of dust levels. The Advisory Committee on Asbestos (1979b) believe that what would have been measured as 2 fibres/ml in 1968 might well be measured as between 4 and 10 fibres/ml ten years later. The implication is that there has been a de facto tightening up of hygienic standards since 1969. However, it also illustrates the problem of providing reliable estimates of dose over a period of time. Knowledge about the dose/response relationship is still restricted by the lack of definitive answers to the following problems (Advisory Committee on Asbestos 1979: 79):

> ...the relative importance in causing asbestos-related disease of factors such as the individual's working lifetime, the length of time that fibres remain in the lungs and exposure to very high peak concentrations; the relative importance of the physical characteristics of fibres (size and shape) and their chemical characteristics (composition and mineral origin) in determining their biological effect.
>
> The biological effect of fibres outside the size range currently specified by the HSE for monitoring purposes.
>
> The feasibility of setting a control limit based on the individual's accumulated exposure.

Pollution in the asbestos industry

Employers in the UK claimed that a 1 fibre/ml standard (on a four-hour time-weighted average) could be met at an additional cost which would put up the price of asbestos-cement products by 1.5 per cent. Increase in the cost of other products would be greater still. There would be further costs due to loss of exports and increased imports and an estimated loss of employment of 5,500 (Table 8.12). If the standard were reduced substantially below the 1 fibre/ml level the costs of production, it was claimed, would increase substantially. Employers believed a 1 fibre/ml standard could be met within eighteen to twenty-four months.

There has, it seems, been an attempt to balance the costs of reducing risks against the dangers of asbestos dust. Unfortunately the Advisory Committee on Asbestos fails to offer convincing estimates of costs involved in meeting different standards. They cost the 1 fibre/ml standard, but suggest that any further reduction would be prohibitively expensive. Estimates of costs are given by the asbestos industry and there is a lack of any independent assessment. The asbestos industry's estimates are based upon an unpublished report (Staniland Hall Report) which was commissioned by the asbestos industry in 1976–7. The report investigated the costs to industry of meeting a 0.5 and 1 fibre standard for chrysotile and amosite. According to this report the effects of the two standards on sales and employment were predicted to be as follows. For a 0.5 fibre standard there would be a £48.5 m. loss of exports, a £82 m. increase in imports and 21,600 people would become unemployed. The 1 fibre standard, however, would lead to considerably less serious consequences. There would be a £14 m. loss of exports, a £16 m. increase in imports, and 5,500 people would lose their jobs. This report was submitted to the Advisory Committee, which accepted the conclusions that a 0.5 fibre standard for chrysotile would be too much of an economic burden. However, the Committee never referred to the Staniland Hall Report nor did it seek to obtain an independent assessment. The recommendation of a 1 fibre standard did little more than confirm what the asbestos industry was willing to comply with. The much longer and somewhat inconclusive reporting of the dose/response relationship was therefore considered of secondary importance to the poorly investigated costs of tighter standards and their enforcement.

Problems of enforcement

It is widely recognized that the 1931 Asbestos Regulations were inadequately enforced. During nearly forty years of its existence only three prosecutions were made by the Factory Inspectorate. One of these prosecutions occurred in 1964 at the Bermondsey factory owned by Central Asbestos. The employers were fined a meagre £170 and 50 p. costs. Later on in 1970 seven workers took Central Asbestos to court for compensation and were awarded £86,469 personal injury damages. Evidence at the ten-day

Table 8.12 Impact on asbestos industries of 1 fibre/ml limit: increases in costs, prices, exports, imports, sales and employment*

	Capital (£ m.)	Current (£ m. p.a.)	Price (%)	Exports (£ m.)	Imports (£ m.)	Home market (%)	Production (%)	Employment (thousands)
Asbestos cement	0.5	0.3	1½	−0.5	—	−10½	−11½	−0.7
Insulation materials	‡	‡	4 †	−3.0	—	−100	−100	−1.0
Friction materials	5.0	4.5	12½ §	−6.0	+14.0	—	−30	−2.7
Gaskets, jointings and packings	0.7	0.4	+3½ §	—	—	—	—	—
Other asbestos textiles millboard and paper	10.0 ¶	3.3	+17	−2.0	+1.5	—	−9	−0.5
Flooring	2.5	1.5	5	−2.5	+0.5	—	−4	−0.6
Paints, adhesives, etc.	—	—	—	—	—	—	—	—
Reinforced plastics	—	—	—	—	—	—	—	—
Total	18.7	10.0		−14.0	+16.0			−5.5

* As compared with 1976
† Would not be incurred because consumers could not comply with standard
§ Including addition to costs of bought-in textiles
¶ Assuming technically feasible
‡ Assuming no change in standard for/availability of asbestos products from abroad, and based on 1976 prices

Source: Asbestos Information Centre.

hearing suggested a persistent and flagrant disregard for the regulations and a failure by the Inspectorate to enforce the Factories Acts (BSSRS, 1979; *The Listener* 2 May 1974). According to interviews with workers from Acre Mill at Hebden Bridge, reported on a BBC 2 *Horizon* programme in January 1975, asbestos regulations were again often abused. The 1931 Regulations banned the use of hand stricklers, but these were used for cleaning carding machines. Regulations banned the use of permeable sacks to carry asbestos, but according to workers hessian sacks were often used.

According to Mendelle, production manager at the Cape Asbestos factory in Barking between 1956 and its closure in 1969, inadequate money was spent on removing the dust completely, but had the company done so it could not have afforded to manufacture asbestos. Here lies the key factor responsible for the attitude of the asbestos industry to the 1931 Regulations. Subsidiary to the interest of keeping costs of manufacture as low as possible are a number of other excuses by management for the failure to deal with the asbestos problem. Management's position on the Acre Mill controversy was that the dangers of asbestos were not fully appreciated, especially until the 1960s. Furthermore, workers were not always conscientious in carrying out the regulations. Respirators, for example, were often uncomfortable and unpleasant to wear. In financial terms, however, the asbestos manufacturers are compelled to keep their costs below or in line with their competitors. Few prosecutions, low fines and a cosy relationship with the Factory Inspectorate has consequently paid dividends in the past.

Publicity has been aimed at allaying fears about contact with asbestos, while much less effort has been made to impress upon workers the real risks of exposure to asbestos dust. Even Mendelle was not informed of the risks when he was appointed Production Manager. Certainly workers are entitled to feel bitter that they were kept so ill-informed.

One of those who knew he was dying from asbestosis expressed the sentiments repeated by countless other victims of the disease (*The Listener* 2 May 1974):

> We've all got to go sometime. It's just that I know I'm going to go sooner than I should be going. I'm not afraid of it. But the knowledge is there, that you're going to go before your time, through no fault of your own. I feel very bitter about it because, if they'd told me of the dangers, I wouldn't be in the position I'm in. Really and truly, they should have told me. They knew.

Selective control of information was certainly in the economic interests of industry, but clearly wasn't in the interest of workers. A weak union and the employment of considerable immigrant labour at Acre Mill made control over information easier. Before 1974 there was no statutory requirement to warn workers of health risks. Workers could only get facts from the inspectors if their employer agreed. If an inspector felt strongly enough to issue warnings against the employer's wishes, he could have been sent to prison. Under the Health and Safety at Work Act 1974 this situation has changed. The employer has a duty to 'provide such information, instruction, training and supervision as is necessary to ensure, so far as is reasonably practicable, the health and safety at work of his employees'. Nevertheless there is still a lack of expertise among workers in recognizing dangers.

Given the imbalance of power between the two sides of industry, could one have expected the Factory Inspectorate to enforce proper control of asbestos dust and inform the workforce of the risks? To answer this question it is necessary to look at the record of the Inspectorate and at the economic context in which the Inspectorate worked.

A detailed study of the Factory Inspectorate's role in controlling asbestos dust was carried out by the Ombudsman in response to complaints of maladministration at Acre Mill (Parliamentary Commission for Administration 1975–6). The factory had been opened in 1939 and was closed in 1970. The report covers a series of visits paid by the Factory Inspectorate to Acre Mill, starting in 1949. Earlier records were not extant. The general conclusions of the report were that the Factory Inspectorate had given insufficient attention to the health risks of Acre Mill, there was an obvious failure by the Inspectorate to appreciate the significance of asbestosis statistics, and there appeared to have been little co-ordination between the chemical branch, which analysed concentrations of asbestos dust, and the engineering branch, which tried to help firms find practical ways of eliminating them. Insufficient pressure was put on the factory to improve its conditions, and when prosecution was advised by the chemical branch on one occasion, the engineering branch advised differently. The Ombudsman was not able to substantiate the widely-held belief among workers that the mill management knew when factory inspectors were coming and ordered them to clean up the factory in advance.

The Ombudsman's report refers on several occasions to the lack of response by management to recommendations made by inspectors, especially between 1965 and 1969. The possibility of bringing prosecutions before the late 1960s was made difficult for a number of reasons. As stated earlier, absolute exclusion of asbestos dust, which was the statutory requirement, proved impractical. There was the problem of accurately determining levels of dust sufficient to support a successful prosecution. Because of the state of medical knowledge, medical inspectors were reluctant to give firm evidence of a link between injury to health and conditions in a factory during a specific period.

During the period before 1970 the Factory Inspectorate saw their role as one of acting through persuasion rather than prosecution. They were clearly understaffed and the derisory fines resulting from prosecution were no incentive to undertake the extra paperwork and suffer the inconvenience involved in bringing a case to court. The fact is that illegally allowing risks to health is regarded as a white-collar crime and punishment has been correspondingly low. There has traditionally been little pay-off from prosecution. Growing concern about asbestos and the adoption of the 1969 Asbestos Regulations have changed the situation a little. While the pay-off from prosecutions is still low there is at least a definite standard upon which to base prosecutions. The unequal balance between industry and its workers has also been modified with the greater determination of workers and the public to reduce health risks from pollution. The Factory Inspectorate no longer operates in a situation where workers are completely ignorant of risks.

In 1971 a three-phase survey began to check the effectiveness of the 1969 Asbestos Regulations and to study the consequences of exposure to varying doses of dust. One phase of the study monitored asbestos-workers in thirteen factories, all of which were member companies of the Asbestosis Research Council in 1975. During the second visit to the factories, 675 representative personal samples were taken in the workers' breathing zone over a four-hour period. Ninety-five per cent were below the current hygiene standard of 2 fibres/ml and 99 per cent were below 4 fibres/ml. In another survey involving 280 representative samples taken from eighteen factories 90 per cent were below 2 fibres/ml and 96 per cent were below 4 fibres/ml (Health and Safety Executive 1977). While there is still some inadequate control in factories dealing with asbestos, the situation among insulation workers and those involved in delagging and demolition is much more serious.

Manufacturing of asbestos products is undertaken in a fixed workplace whereas insulation work and demolition work which brings workers into contact with asbestos typically involve supervision of an itinerant workforce and maintenance of non-stationary equipment. The Advisory Committee on Asbestos admit (1978: 11–12):

> an employer is likely to be reluctant to go to the expense of installing exhaust ventilation to keep down dust levels for a job which will not last years or even months but weeks or days. Often exhaust ventilation is not practicable or at least not reasonably so and, in order to comply with regulations, employers must rely on wetting the material to be handled and on the provision of protective clothing and respiratory protective equipment, but even these precautions may involve transporting the equipment from one site to another and to and from the place where it is stored.

The difficulties facing the Factory Inspectorate in supervising large numbers of small firms, many of which are often short-lived, lead to considerable problems in detecting irresponsible and illegal work with asbestos products. Even when twenty-eight days' notice is given for work involving crocidolite it may be difficult for the Inspector to be present. The Advisory Committee rejected the idea of introducing a licensing system, as there were too many firms involved and there were so many small jobs involving demolition.

The main weakness of the enforcement system remains the derisory fines imposed upon offenders. In 1976, for example, there were twenty-three convictions under the 1969 Asbestos Regulations and the average fine was £45. Consider, for example, a much-publicized case in 1977 of two men employed by a scaffolding company under sub-contract for a firm of insulation contractors (Health and Safety Executive 1978). A warehouse was being stripped of blue asbestos insulation from the roof of the building. The scaffolders were asked to erect their scaffolds over a weekend but they were not provided with protective clothing or respirators, they were unsupervized and worked within a 'snow storm' of dust. Following this serious disregard for the men's health the scaffold contractors and insulation contractors were prosecuted. The scaffold contractors pleaded guilty and were fined £100, the insulation contractors defended the case but were found guilty and fined £150.

Given the risks of such fines it is not surprising that the 1969 Asbestos Regulations are widely abused by demolition firms. The risks of detection and the fines if caught bear little relation to the costs of extra protection and supervision. They bear little relation to the health risks to which workers are subjected by such illegal activities. It is perhaps interesting to reflect that the Advisory Committee on Asbestos was set up following the public revelation of the scandalous lack of dust control enforcement at Acre Mill. Yet the final report of the Advisory Committee gave scant attention to the problems of enforcement.

Conclusions

In this chapter we have seen how pollution control policy in the asbestos industry has changed since the first attempts to regulate the problem in 1931. Changes in standards have been influenced in part by changing views about the hazards of asbestos dust. Understanding the nature of the problem has, however, been considerably influenced by the asbestos industry which has sought in numerous ways to play down the seriousness of the asbestos risk. One of the most important tasks of industry in a capitalist economy is to minimize the costs of production. This includes minimizing expenditure on pollution control when the benefits to the firm of doing so are less than the costs. So long as the health risks are external to the economies of the firm, that is to say the firm has little responsibility to compensate in full the sufferers from asbestos disease, then the task can be seen as one of resisting the imposition of tighter standards.

The asbestos industry's support of image-building research and of public relations exercises weakly disguised as providing information, and its keeping the workforce either uninformed or misinformed, serve the interests of capital. Investigators outside the immediate influence of the asbestos industry have from time to time contributed to a greater understanding of the problem despite some attempts at discrediting findings where there has been an element of uncertainty.

In recent years there has been a response by workers suffering from asbestos-related diseases and those subject to risk. However, the trade union movement in Britain still remains largely apathetic about health hazards at work. Much of the work they should be doing in fighting for better compensation, as well as tighter standards and enforcement, is being undertaken by individuals and voluntary groups such as the Society for the Prevention of Asbestosis and Industrial Diseases. The last exercise in standard-setting was won in the main by the asbestos industry which ensured that any tightening of standards would do little damage to their capacity to make profit. Meanwhile thousands of people will continue to suffer from hazards associated with asbestos.

CHAPTER 9

Conclusions

This book began with a description of some of the conceptual problems to be tackled before effectively dealing with pollution. Pollution itself was defined in terms of its damage to human health, amenity, animals and plants, and to other environmental resources. A distinction was made between natural causes and human causes of damage, pollution being confined to the latter. The notion of damage is crucial to preventing or controlling pollution. For example, it may be more effective to reduce the contact between the pollutant and its target than to reduce the quantities of pollutant. A wide variety of strategies is usually available for reducing damage or indeed for converting pollutants into useful resources. A cost-effective pollution control strategy is defined as one that succeeds in reducing the damage to an agreed objective at the least cost.

Efficiency criteria can be used for determining the objective of the strategy adopted. The most efficient policy is the one that maximizes public benefit bearing in mind the costs of implementation. This type of analysis is usually referred to as cost-benefit analysis. However, it is not always possible or desirable to attach monetary values to costs and benefits. Other types of analysis such as revealed and expressed preference studies, and environmental impact assessment have been used as aids to administration. In practice explicit trade-offs between costs of control and estimated benefits are rarely made by administrative bodies. Studies which form the basis of standard-setting rarely go beyond attempts at assessing the dose/response relationship. Where cost-benefit analysis has been employed, as in the case of asbestos-dust control, the information provided is often totally inadequate.

Pollution control policy has been overwhelmingly concerned with converting harmful substances into harmless but useless substances. Insufficient attention has been paid to the development of preventative strategies. Considerable scope exists when there is a choice of products such as oil and gas instead of coal and nuclear power, and a choice of technology for manufacturing a product – wind and solar power, for instance, might replace conventional means of electricity generation. Modification of processes, redesigning of equipment and recovery of waste products for reuse are additional strategies for preventing pollution. In some instances pollution can be prevented with economic gain to investors. There are, however economic and organizational constraints, and the extent of preventative measures depends upon the social organization of the economy. In China the self-sufficient co-operatives and communes have been particularly successful in integrating pollution control strategies with other social and economic objectives, especially in the field of public health and agriculture.

One of the most intractable problems in pollution control policy is that of acquiring reliable information. Risk assessment based on currently available information can lead to over-confidence about knowledge of ways in which accidents can occur. All the time new combinations of human error and technical failure are being discovered which had not previously been contemplated. The nuclear power station accident at Three Mile Island, Harrisburg, US has been just one of many examples of a surprising series of failures leading to considerable damage.

Uncertainty about risks of pollution and uncertainty about costs of reducing pollution allow enormous scope for distortion in representing what can or should be done. Economic interests have often used uncertainty to seek delay in establishing the extent of risks and, even when the consequences are finally known in more detail, may use uncertainty about costs of control to lobby for weak controls. This was shown to have happened in the case study concerned with the asbestos health hazard.

One of the major dilemmas of our time is to find a way of dealing with unknowable and yet conceivable environmental risks. How can one, for instance, make low probabilities of very serious risks (associated with nuclear power) commensurable with other much better-known but less catastrophic risks (associated with fossil fuels)? This crucial issue in the debate over the virtues of different energy production strategies remains unresolved. Another issue of considerable importance is the problem of access to reliable information. The majority of data on risks and benefits of chemical products, nuclear power, etc. comes from industrial sources where there is both incentive and opportunity for selective control of information.

Political scientists have been interested in different styles of decision-making, the extent of public participation, and upon whether the policy system is devolved or centralized. Typically there is vertical fragmentation of responsibilities between local, regional and central authorities; and horizontal fragmentation between different departments at each level of administration. State intervention has evolved in response to a variety of interests. Planning controls to prevent avoidable pollution consequences, standard-setting, and pollution taxes have been used to settle conflicts of interest between the polluter and the polluted.

Conclusions

Not all pollution problems can be dealt with at a national level alone. In some instances, as in the case of sulphur and carbon oxides, their accumulation may have regional and global consequences. Many pollutants are highly mobile and cross national boundaries. For the most mobile and persistent chemicals such as some pesticides and polychlorinated biphenyls action to control the problem by just a few countries may not be sufficient to give adequate protection to common resources such as seas and the air. Polluting products traded internationally will also have consequences outside the nation of origin. Differences in national policy may interfere with trade. In the EEC, attempts have been made to harmonize pollution control policy of member countries so as to prevent distortions in trade. In some instances standards adopted impose severe costs on industry, as has increasingly been the case for some asbestos products. There is then an economic incentive to manufacture in countries with less stringent controls. Consequently pollution issues are often international in character.

Although there has been considerable progress in recent years towards international agreement there are still considerable barriers to the OECD principles of non-discrimination and equal rights of access. Treaties and conventions often give rise to advisory bodies or commissions with research responsibilites but without the executive powers to settle transfrontier disputes. International regulations covering transportation of products such as oil have grown enormously, but there are still flag of convenience countries which allow substandard operations, and enforcement of regulations is frequently extremely difficult.

One of the main theoretical debates of the last decade has been concerned with whether a policy based on enforcing standards or collecting a pollution tax is likely to be more cost-effective. Much of this debate is unfortunately of little value as it is often assumed that no distortions will occur as a result of political and economic interests. In practice the operation of both standards and taxes is constrained by institutional arrangements and the activities of pressure groups. If one is to explain how some pollutants have been controlled while progress in reducing the damage from other pollutants has been less successful then it is necessary to assess the influence of different styles of administration, public opinion, pressure groups and economic interests. Several case studies have been used to demonstrate the prime importance of economic interests.

Decision-making about pollution problems occurs in a context constrained by legal and administrative requirements, but these reflect and are reinforced by the variety of interests affected. This framework of analysis enables one to distinguish between different categories of pollution control. The first category includes those pollutants whose reduction has been in the direct economic interests of manufacturers and consumers. Often it is profitable to reduce pollution through recycling. Smoke pollution during the twentieth century owes much to the voluntary and advantageous switch by industry and the domestic economy from coal to electricity, gas and oil.

Conclusions

Improvements in the efficiency of production, such as the adoption of fluid-bed combustion, have also led to reduced pollution. A considerable amount of pollution in the past has been a result of inefficient management of resources. Better management has led to less pollution as well as financial advantages.

A second category of pollution problems can be defined in terms not so much of their quantitative reduction, as of their spatial consequences. Changes in residential development and industrial location clearly affect the consequences of pollution. It was shown how decline in smogs in London and elsewhere from the 1880s owes much to the loosening of the urban structure in migration out of central areas by householders and industry, and how their gradual replacement by commercial property, led to a wider dispersion of smoke emissions, and consequent reduction in pollution problems. Once again this type of improvement cannot be attributed to a direct concern about the pollution problem itself but is an indirect consequence of economic development.

A third category of pollution problems is often resolved when there is a conflict between different financial interests. Water pollution control is likely to be most effective when the interests of polluters (domestic and industrial) are in conflict with other interests such as water companies and fisheries which make demands upon the water body. Administrative changes, such as the development of regional water authorities, and the development of controls in the UK, reflect this conflict. In a case study of early industrial air pollution control it was shown how the growing alkali industry came into conflict with wealthy and influential landowners. Common Law actions threatened industrial interests. A means of condensing acidic fumes was developed in response to this threat. Once a means of dealing with pollution existed it was in the interests of both industry and landowners to adopt a statutory standard. Compliance with a standard reduced the threat of an injunction, while landowners benefited from consistent application of control technology.

A fourth category of pollution problems exists where a broad public interest is affected and there is no strong economic interest that can be mobilized to prevent or control a problem. A solution to the problem, however, can be made without seriously threatening financial interests. Much routine pollution control is of this kind. The Clean Air Acts, for example, in the interests of public health enabled local authorities to regulate the height of industrial chimneys.

The fifth and final category of pollution problems is the most serious and difficult to deal with. This occurs when a viable solution to the problem threatens the survival of an established financial interest. The most dangerous and threatening problems occur when there is insufficient pressure to create checks in safeguarding public health and the environment. Less satisfactory solutions have tended to arise when financial interests have been opposed by weak public health and labour groups. This was illustrated in great detail with respect to the asbestos pollution problem. The case does not mean that financial interests are invincible, but does mean that much more effort needs to be exerted within the labour movement if the asbestos disaster and similar problems associated with chemicals in the environment are to be dealt with.

Management of pollution problems can be improved by adopting better techniques of assessment, but constraints exist which limit effectiveness and efficiency. In the final analysis, whether or not pollution will be adequately dealt with depends upon recognition of the problem, upon a means of prevention or control which is economically viable and enforceable, upon the existence of administrative barriers at local, national and international levels, and finally upon whether financial or other interests would gain from its control.

Bibliography

Advisory Committee on Asbestos (1977) *Selected Written Evidence Submitted to the Advisory Committee on Asbestos 1976-7*, Health and Safety Executive, HMSO, London.
Advisory Committee on Asbestos (1978) *Asbestos – Work on Thermal and Acoustic Insulation*, Health and Safety Executive, HMSO, London.
Advisory Committee on Asbestos (1979a) *Final Report of the Advisory Committee*, (Vol. 1), Health and Safety Commission, HMSO, London.
Advisory Committee on Asbestos (1979b) *Final Report of the Advisory Committee*, (Vol. 2), Health and Safety Commission, HMSO, London.
Ahmed, A. K., MacLeod, D. F. and Carmody, J. (1972) Control for asbestos, *Environment*, **14**(10), 16–29.
Alkins, M. H. and Lowe, J. F. (1977) *Pollution Control Costs in Industry: an economic study*, Pergamon Press, Oxford.
Anderson, A. G. (1928) *Report on Atmospheric Pollution 1927-8*, County Borough of Rochdale (National Society for Clean Air Archives, Brighton).
Ashby, E. (1976) Protection of the environment: the human dimension, *Proceedings of the Royal Society of Medicine*, **69**, 721–30.
Ashby, E. (26 January 1978) *Engineers and Politics: a case history*, 22nd Graham Clark Lecture, Council of Engineering Institutions, London.
Ashby, E. and Anderson, M. (1976 and 1977) Studies in the Politics of Environmental Protection: the historical roots of the British Clean Air Act, 1956: I. The awakening of public opinion over industrial smoke, 1843–53, *Interdisciplinary Science Reviews*, **1**, 279–90; 1977: II. The appeal to public opinion over domestic smoke, 1880–92, *Interdisciplinary Science Reviews*, **2**, 9–26; III. The ripening of public opinion, 1898–1952, *Interdisciplinary Science Reviews*, **2**, 190–206.
Ashworth, W. (1954) *The Genesis of Modern British Town Planning*, Routledge & Kegan Paul, London.
Atomic Energy Commission (AEC) (1975) *Reactor Safety Study: an assessment of accident risks in US commercial nuclear power plants* (Norman C. Rasmussen, Study Director), WASH-1400 (NUREC 75/014) US Nuclear Regulatory Commission, Government Printing Office, Washington, D.C.
Auribault, M. (1906) Observations regarding the hygiene and safety of workers in asbestos spinning and weaving mills, *Bulletin de l'inspection du Travail*, 126.

Bibliography

Babich, H., Davis, D. L. and Stotzky, G. (1980) Acid Precipitation: causes and consequences, *Environment*, **22**(4), 6–13 and 40–1

Bachrach, P. and Baratz, M. S. (1963) Decisions and non-decisions: an analytical framework, *American Political Science Review*, **57**, 641–51.

Barde, J., Brown, G. M. Jr, and Buchot, P. F. T. (1979) Water pollution control policies are getting results, *Ambio*, **8**(4), 152–9.

Barltrop, D. (1979) Effect of emissions: lead, *Clean Air*, **9**(3), 77–81.

Barnaby, F. (1980) The controversy over low-level radiation, *Ambio*, **9**, 74–80.

Barnes, R. A. (December 1979) The long-range transport of air pollution: a review of the European experience, *Journal of the Air Pollution Control Association*, 1219–35.

Bates, G. M. (Mar. 1979) River water quality: maintaining the status quo?, *Journal of Planning and Environmental Law*, **5**, 152–8.

Beaver Committee (1953) *Committee on Air Pollution Interim Report*, Cmd 9011, HMSO, London.

Beaver Committee (1954) *Report of the Committee on Air Pollution*, Cmd 1322, HMSO, London.

Beaver, S. H. (June 1964) The potteries, a study in the evolution of a cultural landscape, *Transactions of the Institution of British Geographers*, **34**, 1–31.

Becker, G. S. (1968) Crime and punishment: an economic analysis, *Journal of Political Economy*, **76**, 169–217.

Beckerman, W. (1973) Pollution control: who should pay? How? How much?, in Clayton, K. C. and Chilver, R. C. (eds) *Pollution Abatement*, David and Charles, Newton Abbot.

Beckerman, W. (1975) *Pricing for Pollution*, Hobart Paper No. 66, Institute of Economic Affairs, London.

Behr, P. (1978) Controlling chemical hazards, *Environment*, **20**, 25–9.

Benn, A. (1979) Science, technology and liberty, *Vole*, **2**, 7–11.

Bennett, G. (1979) Pollution control in England and Wales: a review, *Environmental Policy and Law*, **5**, 93–9.

Bisset, R. (Jul. 1978) Environmental impact analysis, *Royal Anthropological Institute News*, **26**, 1–4.

Blacker, S. M., Ott, W. R. and Stanley, T. W. (1977) Measurement and the law: monitoring for compliance with Clean Air Amendments of 1970, *International Journal of Environmental Studies*, **11**, 169–85.

Bohm, P. and Henry, C. (1979) Cost-benefit analysis and environmental effects, *Ambio*, **8**, 18–24.

Bonaccorsi, A., Fanelli, R. and Tognoni. G. (1978) In the wake of Seveso, *Ambio*, **7**, 234–9.

Booth, H. and Green, A. (1976) The European Community Environmental Programme and United Kingdom law, *European Law Review*, **1**, 444–63.

Borch, K. (1968) *The Economics of Uncertainty*, Princeton University Press, Princeton, N.J.

Bouveng, H. O. (19 April 1980) Pollution control in Swedish law and practice – an industrial point of view, *Chemistry and Industry*, **8**, 331–5.

Bove, F. (1979) The government discovers solar energy, *Science for the People*, **11**(2), 7–12.

Brown, E. H. P. (1959) *The Growth of British Industrial Relations*, Macmillan, London.

Bibliography

Brown, M., Fitzgerald, P., Goodenough, M., Hollenbach, D., Pector, J., Schwartz, D. and Swartz, J. (May 1976) Nuclear power who needs it? *Science for the People*, **8**(3), 4–12.

BSSRS (1979) *Asbestos killer dust*, BSSRS Publications, London.

Buchanan, J. M. and Tullock, G. (May 1975) Polluters' profits and political response: direct controls versus taxes, *American Economic Review*, 141–2.

Buchanan, W. D. (1979) Asbestos-related disease, Part I Introduction, in Michaels, L. and Chissick, S. S. (eds) *Asbestos. Volume I: Properties, applications and hazards*, John Wiley, Chichester, 395–408.

Bugler, J. (1972) *Polluting Britain*, Penguin, Harmondsworth.

Bugler, J. (1975) Bedfordshire Brick, in Smith, P. J. (ed.) *The Politics of Physical Resources*, Penguin, Harmondsworth.

Bugler, J. (Sep. 1978) Pollution standards plunge, *Vole*, **1**(12), 3–4.

Bugler, J. (28 Oct. 1979) The nuclear kingdom, *Observer Magazine*, 31–41.

Bulletin of the European Communities (1976) (Supplement 6/76) *Environment Programme 1977–1981*, Commission of the European Communities, Brussels.

Burge, E. J. (28 July 1978) Energy research and social policy, *Times Higher Education Supplement*, 11.

Burrows, P. (1974) Pricing versus regulation for environmental protection, in Culyer, A. J. (ed.) *Economic Policies and Social Goals: aspects of public choice* (York Studies in Economics, No. 1), Martin Robertson, London, 273–83.

Cairns, W. (1978) The Flotta EIA study, *Built Environment*, **4**, 129–33.

Carpentier, M. (22 April 1977) A review of the scope and progress of the Commission's programme to date. Paper for a seminar entitled 'The European Community's Environment Policy' held in London and sponsored by the European Commission. Copy from National Council of Social Service, London.

Carson, W. G. (1971) White-collar crime and enforcement of factory legislation, in Carson, W. G. and Wiles, P. (eds) *The Sociology of Crime and Delinquency in Britain*, Martin Robertson, London.

Castleman, B. I. (1979) *Impending Proliferation of Asbestos*, paper presented at the conference Exportation of Hazardous Industries to Developing Countries, New York, 2 November 1979.

Central Advisory Water Committee (1971) *The Future Management of Water in England and Wales*, HMSO, London.

Central Unit on Environmental Pollution (1974) *The Monitoring of the Environment in the United Kingdom*, Pollution Paper No. 1, HMSO, London.

Central Unit on Environmental Pollution (1976a) *Accidental Oil Pollution of the Sea*, Pollution Paper No. 8, HMSO, London.

Central Unit on Environmental Pollution (1976b) *Pollution Control in Great Britain: How it Works*, Pollution Paper No. 9, HMSO, London.

Chapman, P. (1976) *Fuel's Paradise*, Penguin, Harmondsworth.

Chissick, S. S. (1979) Attitudes to asbestos, in Michaels, L. and Chissick, S. S. (eds) *Asbestos. Volume 1: Properties, Applications and Hazards*, John Wiley, Chichester, 115–69.

Chugh, L. C., Hanemann, M. and Mahapatra, S. (1978) Impact of pollution control regulations on the market risk of securities in the U.S., *Journal of Economic Studies*, **5**(1), 64–70.

Clarke, R. (May 1972) Soft Technology: blueprint for a research community, *Undercurrents*, **2**.

Clarke, R. (1973) The pressing need for alternative technology, *Impact of Science on Society*, **23**(4), 257–71.
Clifton, R. A. (1974) *Asbestos*, in *Bureau of Mines Minerals Yearbook 1974*, U.S. Government Printing Office, Washington, D.C.
Coase, R. (1960) The problem of social cost, *The Journal of Law and Economics*, **3**, 1–44.
Cohen, J. B. (1896) The character and extent of air pollution in Leeds. A lecture given to the Leeds Philosophical Society on 3 March 1896 (National Society for Clean Air Archives, Brighton).
Commission of the European Communities (1977a) *Public Health Risks of Exposure to Asbestos*, Pergamon Press, Oxford.
Commission of the European Communities (1977b) *State of the Environment*, First Report, EEC, Brussels.
Common, M. S. (1977) A note on the uses of taxes to control pollution, *Scandinavian Journal of Economics*, **79**, 347.
Council on Environmental Quality (1975) *Environmental Quality: sixth annual report*, U.S. Government Printing Office, Washington, D.C.
Council on Environmental Quality (1976) *Environmental Quality: seventh annual report*, Government Printing Office, Washington, D.C.
Council on Environmental Quality (1977) *Environmental Quality: eighth annual report*, U.S. Government Printing Office, Washington, D.C.
Council on Environmental Quality (1978) *Environmental Quality: ninth annual report*, U.S. Government Printing Office, Washington, D.C.
Council on Environmental Quality (1979) *Environmental Quality: tenth annual report*, U.S. Government Printing Office, Washington, D.C.
Crenson, M. A. (1971) *The Un-politics of Air Pollution: a study of non-decision-making in the cities*, Johns Hopkins University Press, Baltimore.
Department of the Environment (1971) *108th Annual Report on the Alkali etc. Works*, HMSO, London.
Department of the Environment (1973) *110th Annual Report on the Alkali etc. Works*, HMSO, London.
Department of the Environment (1974) *111th Annual Report on the Alkali etc. Works*, HMSO, London.
Department of the Environment (1978) *Digest of Environmental Pollution Statistics, No. 1*, HMSO, London.
Department of the Environment (1980) *Digest of Environmental Pollution Statistics, No. 2*, HMSO, London.
Dickson, D. (1974) *Alternative Technology and the Politics of Technical Change*, Fontana, Glasgow.
Downs, A. (1972) Up and down with ecology – the 'issue-attention cycle', *The Public Interest*, **28**, 38–50.
Draggan, S. (1978) TSCA: the US attempt to control toxic chemicals in the environment, *Ambio*, **7**, 260–2.
DSIR (1929) *Report on the Investigation of Atmospheric Pollution Ending 31 March 1929*, HMSO, London.
DSIR (1937) *The Investigation of Atmospheric Pollution. Report on Observations in the Year Ended 31 March 1936*, 22nd Report, HMSO, London.
DSIR (1949) *26th Report of the Investigation of Atmospheric Air Pollution – Observations on the 5 Years Ended 31 March 1944*, HMSO, London.

Bibliography

DSIR (1955) *The Investigation of Atmospheric Pollution. Report on Observations in the 10 Years Ended 31 March 1954*, HMSO, London.

DSIR (1962) *The Investigation of Atmospheric Pollution 1958-63*, Warren Spring Laboratory, HMSO, London.

Durkheim, E. (1950) *Rules of Sociological Method*, eighth edn, The Free Press, Glenco, Illinois.

Edwards, M. (1977) *The Ideological Function of Cost-benefit Analysis in Planning*, Discussion Paper No 25, University College, London.

Elkington, J. (2 June 1977) The ecology of tomorrow's water, *New Scientist*, **74**, 524-5.

Elkington, J. (1980) *The Ecology of Tomorrow's World – Industry's Environment*, Associated Business Press, London.

Elliott, D. (1978) *The Politics of Nuclear Power*, Pluto Press, London.

Elliott, D. (1979) Decentralising AT, *Undercurrents*, **35**, 21-3.

Epstein, S. (1979) Polluted data, *The Ecologist*, **9**, 264-8.

Fanning, O. (1975) *Man and His Environment: citizen action*, Harper and Row, New York.

Fergusson, W. C. (1974) Plastics, their contribution to society and considerations of their disposal, in Staudinger, J. J. P. (ed.) *Plastics and the Environment*, Hutchinson, London.

Fernie, J. (1980) *A Geography of Energy in the United Kingdom*, Longman, London.

Fischoff, B., Hohenemser, C., Kasperson, R. E. and Kates, R. W. (1978a) Handling hazards, *Environment*, **20**(7) 16-20 and 32-7.

Fischoff, B., Slovic, P. and Lichtenstein, S. (1979) Weighing the risks, *Environment*, **21**(4), 17-20 and 32-8.

Fischoff, B., Slovic, P., Lichtenstein, S., Read, S. and Combs, B. (1978b) How safe is safe enough? *Policy Sciences*, **9**, 127-52.

Flood, M. and Grove-White, R. (1976) *Nuclear Prospects: a comment on the individual, the State and nuclear power*, Friends of the Earth, in association with the Council for the Protection of Rural England and the National Council for Civil Liberties, London.

Flowers, B. (Autumn 1977) HMPI in wonderland and vexations, *Clean Air*, **7**, 7.

Flowers, B. (1979) Energy and the environment – the public debate, *Clean Air*, **9**(2) 48-57.

Frankel, M. (Spring 1974) The Alkali Inspectorate: the control of industrial air pollution, *Social Audit*, **1**(4), supplement, 1-48.

Frankel, M. (1978) *The Social Audit Pollution Handbook. How to Assess Environmental and Workplace Pollution*, Macmillan, London.

Frederickson, H. G. and Magnas, H. (1972) Comparing attitudes towards water pollution in Syracuse, in Thompson, D. L. (ed.) *Politics, Policy and Natural Resources*, The Free Press, New York, 264-80.

Freeman III, A. M. (1971) *The Economics of Pollution Control and Environmental Quality*, General Learning Press, New York City.

Freeman, L. (November 1977) Trade effluents: the charges and the consequences, *Water*, 13-16.

Fremlin, J. (1980) Health risks from low-level radiation, *Ambio*, **9**, 60-5.

Gardiner, J. (19 April 1980) Monitoring and control of discharges to tidal waters, *Chemistry and Industry*, **8**, 307-12.

The Glass Container Industry and the Environmental Debate (1975) Glass Manufacturers Federation, London, 24–5.
Goldman, M. (1972) *The Spoils of Progress: environmental pollution in the Soviet Union*, MIT, Cambridge, Mass.
Goodin, R. E. (1978) Uncertainty as an excuse for cheating our children: the case of nuclear wastes, *Policy Sciences*, **10**, 25–43.
Greenberg, M. and Lloyd Davies, T. A. (1974) Mesotheliomia register 1967–8, *British Journal of Industrial Medicine*, **31**, 91.
Greene, W. (1978) A conversation with the New Alchemists, *Environment*, **20**(10) 25–8.
Grove-White, R. (20 May 1976) U.K. water policy and public scrutiny, *New Scientist*, **70**, 424.
Gunningham, N. (1974) *Pollution, Social Interest and the Law*, Martin Robertson, London.
Gustafson, T. (1978) The new Soviet environmental programme: do the Soviets really mean business? *Public Policy*, **26**, 455–76.
Hägerhäll, B. (1980) International co-operation to protect the Baltic, *Ambio*, **9**, 183–6.
Hall, I. M. (1976) *Community Action Versus Pollution: a study of a residents' group in a Welsh urban area*, University of Wales Press, Cardiff.
Hammons, A. S. and Huff, J. E. (1974) Asbestos: world concern, involvement and culpability, *International Journal of Environmental Studies*, **6**, 247–52.
Hanlon, J. (27 October 1977) Visiting windmills in Wales, *New Scientist*, **76**, 216–18.
Hardie, D. W. F. (1950) *A History of the Chemical Industry in Widnes*, ICI, London.
Harper, P. (1973) In search of allies for the soft technologies, *Impact of Science on Society*, **23**(4), 287–305.
Harper, P. (1976) Economics of autonomy, in: Boyle, G., Harper, P. and the editors of *Undercurrents, Radical Technology*, Wildwood House, London, 156–62.
Hartley, D. A. (1972) Inspectorates in British Central Government, *Public Administration*, **50**, 447–66.
Hawkesworth, D. L. and Rose, F. (1970) Qualitative scale for estimating sulphur dioxide air pollution in England and Wales using epiphytic lichens, *Nature*, **227**, 145–8.
Health and Safety Executive (1977) *Industry and Services 1975*, HMSO, London.
Health and Safety Executive (1978) *Manufacturing and Service Industries 1977*, HMSO, London.
Herbert, J. H., Swanson, C. and Reddy, P. (1979) A risky business: energy production and the Inhaber Report, *Ambio*, **20**, 28–33.
Holdgate, M. W. (1979) *A Perspective of Environmental Pollution*, Cambridge University Press, Cambridge.
Holloway, T. (1978) Back from the dead. The restoration of the River Thames, *Environment*, **20**(5), 6–11.
Horn, J. S. (1969) *Away with All Pests*, Hamlyn, London.
Howard, R. A., Matheson, J. E. and Owen, D. L. (1978) The value of life and nuclear design, in: Okrent, D. and Cramer, E. (eds) *Probabilistic Analysis of Nuclear Reactor Safety*, American Nuclear Society Publication, La Grange Park, III, 1978; IV, 2.1 to IV 2.9. Quoted in Fischoff *et al.* (1979: 17).

Bibliography

Howard, S., Kimber, L. J., Lewinsohn, H. C., Peto, J. and Doll, R. (1976) *A Mortality Study among Workers in an English Asbestos Factory*, Oxford University Press, Oxford.

Howe, G. M. (1972) *Man, Environment and Disease in Britain. A Medical Geography of Britain Through the Ages*, David and Charles, Newton Abbot.

Howell, D. (Autumn 1976) The future of pollution control, *Clean Air*, **9**.

Industrial Air Pollution 1975 (1977) Health and Safety Executive, HMSO, London.

Industrial Air Pollution 1976 (1978) Health and Safety Executive, HMSO, London.

Industrial Air Pollution 1977 (1979) Health and Safety Executive, HMSO, London.

Industrial Air Pollution 1978 (1980) Health and Safety Executive, HMSO, London.

Inhaber, H. (18 May 1978) Is solar power more dangerous than nuclear? *New Scientist*, **78**, 444–6.

Ireland, F. E. (Summer 1977) Reflections of an alkali inspector, *Clean Air*, **7**(25), 4–9.

Jackson, C. I. (1971) The dimension of international pollution, *Oregon Law Review*, **50**, 223–58.

Jevons, W. S. (1906) *The Coal Question*, third edn, revised and edited by Flux, A. W. 1965, A. M. Kelley, New York.

Johnson, S. (1977) An exposition of the Commission's policies, in *The European Community's Environmental Policy*, European Commission sponsored seminar in London on 22 April 1977.

Johnson, S. P. (1978) Developments in the EEC for control of new chemical substances, *Ambio*, **7**, 267–70.

Jones, G. G. (15 January 1977) Legislation and the role of the alkali inspectorate, *Chemistry and Industry*, **2**, 54–8.

Joyce, C. (11 September 1980) Industrial China's expensive dirt, *New Scientist*, **88**, 772–5.

Kelley, D. R. (1976) Environmental policy-making in the USSR: the role of industrial and environmental interest groups, *Soviet Studies*, **28**, 570–89.

Kelley, D. R., Stunkel, K. R. and Wescott, R. R. (1975) The politics of the environment, in: Milbraith, L. W. and Inscho, F. R. (eds) *The Politics of Environmental Policy*, Sage, London, 115–34.

Kelley, D. R., Stunkel, R., Wescott, R. R. (1976) *The Economic Superpowers and the Environment – the United States, the Soviet Union, and Japan*, W. H. Freeman, San Francisco.

Kimber, R. and Richardson, J. J. (1974) *Campaigning for the Environment*, Routledge and Kegan Paul, London.

Kinnersley, P. (1980) Inter-city drain, *Undercurrents*, **38**, 19–20.

Kneese, A. V., Ayers, R. U. and D'Arge, R. C. (1970) *Economics and the Environment*, Johns Hopkins University Press, Baltimore.

Kneese, A. V. and Schultze, C. L. (1975) *Pollution, Prices and Public Policy*, The Brookings Institution, Washington, D.C.

Kotelchuck, D. (Sep. 1975) Asbestos – science for sale, *Science for the People*, 8–16.

Lawther, P. J. (1973) Air pollution and its effects on man, in: Clayton, K. M. and Chilver, R. C. (eds) *Pollution Abatement*, David and Charles, Newton Abbot, 39–60.

Leach, G. (1979) *A Low Energy Strategy for the U.K.*, IIED, London.

Lerch, I. (3 January 1980) Risk and fear, *New Scientist*, **85**, 8–11.
Littlejohn, H. (1897) *Report on the Causes and Prevention of Smoke from Manufacturing Chimneys*, City of Sheffield (National Society for Clean Air Archives, Brighton).
Liu, B. (1979) The Costs of air quality deterioration and benefits of air pollution control, *American Journal of Economics and Sociology*, **38**, 187–95.
Lovins, A. B. (1977) *Soft Energy Paths: toward a durable peace*, Penguin, Harmondsworth.
Lundqvist, L. J. (1979) Who is winning the race for clean air? An evaluation of the impacts of the U.S. and Swedish approaches to air pollution, *Ambio*, **8**, 144–51.
Lutz, II, R. E. (1976) The laws of environmental management: a comparative study, *The American Journal of Comparative Law*, **24**, 447–520.
McIntosh, P. (Nov. 1978) Water pollution, charging systems and the EEC, *Water*, **23**, 2–6.
McKean, R. N. (1980) Enforcement costs in environmental and safety regulation, *Policy Analysis*, **6**, 269–89.
Mackeown, T. and Record, R. G. (1962) Reasons for the decline of mortality in England and Wales during the nineteenth century, *Population Studies*, **16**, 94–122.
McKillop, A. (Mar./Apr. 1978) A Tsunamai of wave power, *New Ecologist*, **2**, 42–4.
McLeod, R. M. (1965) The Alkali Acts Administration, 1863–84. Re-emergence of the civil scientist, *Victorian Studies*, **9**, 85–112.
McLoughlin, J. (1976) *The Law and Practice Relating to Pollution Control in the United Kingdom*, Graham and Trotman, London.
Maddox, J. (1975) *Beyond the Energy Crisis*, Hutchinson, London.
Majone, G. (1976a) Standard setting and the theory of institutional choice: the case of pollution control, *Policy and Politics*, **4**, 35–51.
Majone, G. (1976b) Choice among policy instruments for pollution control, *Policy Analysis*, **2**, 589–613.
Malcolm, C. V. (Spring 1977) Smokeless zones – the history of their development: Part II. – the development of smokeless zones 1934–1958, *Clean Air*, **7**, 4–9.
Mandl, V. (May 1979) EEC approach to water pollution, *Water*, **26**, 2–4.
Marquand, J. (2 October 1976) An economist's view of water pollution charges as regulatory instruments, *Chemistry and Industry*, **19**, 835–9.
Marquand, J. (1977) *Economic Information for Environmental (Anti-Pollution) Policy*, Government Economic Service Occasional Paper No. 13, HMSO, London.
Martin, C. (1 March 1980) Insulating materials and lung disease, *Chemistry and Industry*, 172.
Martindale, R. (1976) How should industry view pollution charges? *CBI Review*, **21**, 11–20.
Martindale, (17 March 1979) Charging for direct discharges, *Chemistry and Industry*, 195–8.
Mellanby, K. (1975) *Can Britain Feed Itself?* Merlin Press, London.
Mitchell, B. (1979) *Geography and Resource Analysis*, Longman, London.
Mitchell, B. R. and Deane, P. (1962) *Abstract of British Historical Statistics*, Cambridge University Press, Cambridge.

Bibliography

Molotch, H. and Lester, M. (1973) Accidents, scandals and routines: resources for insurgent methodology, *The Insurgent Sociologist*, **3**, 1–11.

Müller, F. G. (1979) Divide-up to clean-up, *Environmental Policy and Law*, **5**, 13–15.

Musgrove, P. (8 June 1978), *New Scientist*, **78**, 694.

Nagel, S. (1978) Parliamentary action on the *Amoco Cadiz, Environmental Policy and Law*, **4**, 167–9.

Nelkin, D. (1975) The political impact of technical expertise, *Social Studies of Science*, **5**, 35–54.

Nomura, Y. (1975–6) Pollution-related injury in Japan, *Environmental Policy and Law*, **1**, 179–83.

OECD Environment Directorate (1974), *The Polluter-Pays Principle: note on the implementation of the polluter-pays principle*, OECD, Paris.

OECD (1976b) *Economic Measurement of Environmental Damage*, OECD, Paris.

OECD (1977a) *OECD and the Environment*, OECD, Paris.

O'Riordan, T. (1976) *Environmentalism*, Pion, London.

O'Riordan, T. (1979) The scope of environmental risk management, *Ambio*, **8**, 260–4.

Orleans, L. A. (1975–6) China's environomics: backing into ecological leadership, *Environmental Policy and Law*, **1**, 189–93 and **2**, 28–31.

Otter, R. (5 May 1979) Environmental quality objectives, *Chemistry and Industry*, **9**, 302–5.

Parker, D. J. and Penning-Rowsell, C. (1980), *Water Planning in Britain*, George Allen and Unwin, London.

Parliamentary Commission for Administration (1975–6) *Third Report*, House of Commons Paper No. 259, 189–211, HMSO, London.

Pearce, D. (4 December 1976a) Forecasting the supply and demand for waste paper, *Materials Reclamation Weekly*, 22–5.

Pearce, D. (11 December 1976b) Forecasting the supply and demand for waste paper, *Materials Reclamation Weekly*, 20–2.

Pearce, D. (1976c) *Environmental Economics*, Longman, London.

Peto, R. (27 March 1980) Distorting the epidemiology of cancer: the need for a more balanced overview, *Nature*, **284**, 297–300.

Pochin, E. E. (1973) Hazards from radiation, in: Clayton, K. M. and Chilver, R. C. (eds) *Pollution Abatement*, David and Charles, Newton Abbot, 32–9.

Port, G. N. J. (1978) Information systems for the control of toxic chemicals in the environment, *Ambio*, **7**, 271–4.

Posner, R. A. (1972) *The Economic Analysis of Law*, Little, Brown and Co., Boston.

Renshaw, D. C. (1978) How and why legislation is produced, WPC Annual Conference, 1978, Paper No. 3.

Report of a River Pollution Survey of England and Wales 1970, (1971), (Vol. 1), Department of the Environment, HMSO, London.

Richardson, J. J. (1979) Agency behaviour: the case of pollution control in Sweden, *Public Administration*, **57**, 471–82.

Ridley Report (1952) *Report of the Committee on National Policy for the Use of Fuel and Power Resources*, Cmd 8647, HMSO, London.

Rieser, R. (1973) The territorial illusion and behavioural sink: critical notes on behavioural geography, *Antipode*, **5**, 52–7.

Roos, L. L. and Roos, N. P. (1972) Pollution regulation and evaluation, *Law and Society Review*, **6**, 509–29.
Rosencranz, A. (1980) The problem of transboundary pollution, *Environment*, **22**(5), 15–20.
Rosencranz, A. and Wetstone, G. (1980) Acid precipitation: national and international responses, *Environment*, **22**(5), 6–8.
The Rotheram Study (1973), Department of Environment, London.
Royal Commission on Environmental Pollution (1976a) *Air Pollution Control: an Integrated Approach*, Fifth Report, Cmd 6371, HMSO, London.
Royal Commission on Environmental Pollution (1976b) *Nuclear Power and the Environment*, Sixth Report, Cmd 6618, HMSO, London.
Royston, M. G. (1979) *Pollution Prevention Pays*, Pergamon Press, Oxford.
Sage, B. (19 April 1979) Disaster at Sullom Voe, *New Scientist*, **82**, 183–4.
Sandbach, F. R. (Mar. 1977a) Threat of river pollution to future water supplies, *Environmental Policy and Law*, **3**(1), 32–4.
Sandbach, F. R. (April 1977b) Have rivers now reached the age of consent? *Municipal Engineering*, **154**, 430–4.
Sandbach, F. R. (1977c) Public participation and the Control of Pollution Act 1974, *New Law Journal*, **127**, 652–4.
Sandbach, F. R. (1977d) Farewell to the god of plague – the control of schistosomiasis in China, *Social Science and Medicine*, **11**, 27–33.
Sandbach, F. R. (1979) Economics of pollution control, in: Lenihan, J. and Fletcher, W. W. (eds) *Economics and the Environment*, Environment and Man Series (Vol. 10), Blackie, Glasgow.
Sandbach, F. R. (1980) *Environment, Ideology and Policy*, Basil Blackwell, Oxford.
Sanderson, J. B. (1974) The National Smoke Abatement Society and the Clean Air Act (1956), in: Kimber, R. and Richardson, J. J. (eds) *Campaigning for the Environment*, Routledge and Kegan Paul, London, 27–44.
Saunders, P. J. W. (1976) *The Estimation of Pollution Damage*, Manchester University Press, Manchester.
Scarrow, H. A. (1972) The impact of British domestic air pollution legislation, *British Journal of Political Science*, **2**, 261–82.
Select Committee on Noxious Vapours (1862) *Report* PP (486), 7.
Self, P. (1975) *Econocrats and the Policy Process, the Politics and Philosophy of Cost-Benefit Analysis*, Macmillan, London.
Selikoff, I. J. and Lee, D. H. K. (1978) *Asbestos and Disease*, Academic Press, London.
Sharp, P. G. (Summer 1976) National and international factors influencing changes in the policy of air pollution control, *Clean Air*, 6–10.
Sigurdson, J. (1972) China: re-cycling that pays, *Läkartidningen*, **69**, 2837–41.
Simon Report (1946) *Domestic Fuel Policy*, Cmd 6762, HMSO, London.
Sinclair, C. (1972) *A Cost-Effectiveness Approach to Industrial Safety*, HMSO, London.
Singer, A. (24 October 1979) Brushing aside the deadly dust, *The Guardian*, 16.
Slovic, P., Fischoff, B. and Lichtenstein, S. (1979) Rating the risks, *Environment*, **21**(3), 14–20 and 36–9.
Smets, H. (1976) Local actions to combat transfrontier pollution, *Ambio*, **5**, 164–8.
Sørensen, B. (1979) Nuclear power: the answer that became a question, *Ambio*, **7**, 10–7.

Steck, H. J. (1975) Private influence on environmental policy: the case of the National Industrial Pollution Control Council, *Environmental Law*, **5**, 241–81.

Stigler, G. J. (1970) The optimum enforcement of laws, *Journal of Political Economy*, **78**, 526–36.

Storey, D. J. (1977) A socio-economic approach to water pollution law enforcement in the United Kingdom, *International Journal of Social Economics*, **4**, 207–24.

Storey, D. J. and Elliott, D. J. (7 May 1977) An effluent charging scheme for the river Tees, *Chemistry and Industry*, **33**, 5–8.

Strigini, P. and Torriani-Gorini, A. (1977) Seveso: Zona Infestata, *Science for the People*, **9**, 8–16.

Taylor, J. S. (1929) *Smoke and Health*, Manchester & District Regional Smoke Abatement Committee, National Smoke Abatement Society, Brighton.

Thairs, E. P. (Winter 1977) British industry and pollution control in the EEC, *CBI Review*, **23**, 27–35.

Thomas, C. (1977) *The Paper Chain*, Earth Resources Research, London.

Todd, J. (Spring 1971) A modest proposal, *The New Alchemy Institute Bulletin*.

Todd, J. (1976) Pioneering for the 21st century: a new alchemist's perspective, *The Ecologist*, **6**(7), 252–7.

Transport Policy (1976) *Consultation Document (Vol. 2)*, HMSO, London.

UNEP Annual Report (1978) Chemicals and the Environment, reprinted in *Ambio*, **7**, 240–3.

Van Buren, E. A. (1980) Biogas beyond China: first international training program for developing countries, *Ambio*, **9**(1), 10–15.

Victor, P. A. (1972) *Pollution: economy and the environment*, Allen and Unwin, London.

Wainwright, M. (1980) Man-made omissions of sulphur and the soil, *International Journal of Environmental Studies*, **14**, 279–88.

Walter, I. (1975) *International Economics of Pollution*, Macmillan, London.

Walter, R. and Storper, M. (1978) Erosion of the Clean Air Act of 1970: a study in the failure of government regulatory planning, *Environmental Affairs*, **7**, 189–257.

War on Waste: a policy for reclamation (1974) Cmd 5727, HMSO, London.

Wardley-Smith, J. (1979) Sources of pollution, in Wardley-Smith, J. (ed.) *The Prevention of Oil Pollution*, Graham and Trotman, London, 1–16.

Waste Management Advisory Council (1976) *Report on Waste Paper Collection by Local Authorities*, Paper No. 2. Department of the Environment, HMSO, London.

Wathern, P. (1976) The role of impact statements in environmental planning in Britain, *International Journal of Environmental Studies*, **9**, 165–8.

Watson, W. D. (1978) *Costs and Benefits of Water Pollution*, Discussion Paper D–18, Division of Renewable Resources, Resources for the Future, Washington, D.C.

Wenner, L. M. (1978) Pollution control: implementation alternatives, *Policy Analysis*, **4**(1), 47–65.

West, R. and Foot, P. (1975) Anglesey: aluminium and oil, in Smith, P. J. (ed.) *The Politics of Physical Resources*, Penguin, Harmondsworth, 202–32.

Westergard, J. and Resler, H. (1976) *Class in a Capitalist Society*, Penguin, Harmondsworth.

Wetstone, G. S. (1980) The need for a new regularity approach, *Environment*, **22**(5) 9-14 and 40-3.
Wharan, A. (13 March 1975) Problem of international law, *New Law Journal*, **125**, 268-9.
White, L. J. (1976a) Effluent charges as a faster means of achieving pollution abatement, *Public Policy*, **24**, 111-25.
White, L. J. (1976b) American automotive emissions control policy: a review of the reviews, *Journal of Environmental Economics and Management*, **2**, 231-46.
Wilde, P. J. (Winter 1978) EEC approach to air pollution control – the government viewpoint, *Clean Air*, 14-22.
Wood, C. (15 June 1978) DoE bars Cheshire air standards, *New Scientist*, **78**, 738-9.
Wood, C. M., Lee, N., Luker, J. A. and Saunders, P. J. W. (1974) *The Geography of Pollution – A Study of Greater Manchester*, Manchester University Press, Manchester.
Working Party on Sewage Disposal (1970) *Taken for Granted*, Ministry of Housing and Local Government, HMSO, London.
World Health Organization (1972) *Air Quality Criteria and Guides for Urban Air Pollutants. Report of a WHO Expert Committee*, Technical Report Series, No. 506, World Health Organization, Geneva.
World Health Organization (1977) *Asbestos*, International Agency for Research on Cancer. IARC Monographs on the evaluation of carcinogenic risk of chemicals to man, (Vol. 14), World Health Organization, Geneva.
Yishai, Y. (1979) Environment and developmment: the case of Israel, *International Journal of Environmental Studies*, **14**, 205-16.

Index

Advisory Committee on Asbestos, 142, 143, 145, 147, 148, 152, 153
Alkali Act 1863, 115–18
Alkali and Clean Air Inspectorate, 69, 86, 91, 94–100, 117, 121, 122
Alternative technology, 6, 31, 42–7
Antagonism, 13
Asbestos, 25, 26, 124–53
Asbestos Industry Regulations 1931, 139, 140, 148, 150
Asbestos Information Centre, 142–4
Asbestos Regulations 1969, 128, 140–2, 152, 153
Asbestosis, 131–4, 138–41
Asbestosis Research Council, 135, 142, 143

Best practicable means, 75, 94, 95, 97, 98, 101
Boundaries Water Treaty, 52

Cancer, xiii, 26, 28, 29, 58, 125, 131–4, 138–41
Carcinogens, 13, 14, 34, 35
Central Unit on Environmental Pollution, 62
Clean Air Act 1956, 13, 15, 49, 63, 78, 104, 106, 112, 114, 115
Clean Air Act 1970, 1, 16, 82, 121
Common law, 79
Control of Pollution Act 1974, 85, 93, 94, 99, 103
Cost-benefit analysis, 19, 20, 77, 148
Cost-effectiveness, 5
Council on Environmental Quality, 62

Department of the Environment, 62
Dose/response relationship, 11–14, 16, 147

Economic efficiency, 10–11
Enforcement, 84, 88–94, 96, 148, 150–3
Environmental impact assessment, 20
Environmental Protection Agency, 11, 61, 66, 67, 82, 92
EEC Environment Programme, 74, 76
Event-tree, 27, 28
Export of pollution, 58, 59
Expressed preferences, 18–20

Fault-tree, 27, 28
Federal Water Pollution Control Act Amendments 1972, 1, 16, 92

Helsinki Convention 1974, 52

Inter-Governmental Maritime Consultative Organization, 54, 55

London smog of 1952, 3, 103, 107, 114

Merewether Report 1930, 131, 139
Mesothelioma, 132–4, 138–41

National Environmental Policy Act 1969, 20, 67
Nordic Environmental Protection Agreement 1974, 50
Nuclear power, 27–35, 42, 65

Oil pollution, 53–5
Optimum level of pollution control, 77–80

Index

Pathway analysis, 8
Pluralism, 102–4, 118, 119, 122, 123
Pollution control subsidies, 77, 78
Pollution tax, 77, 79–89, 101
Public opinion, 103, 104–6
Public participation, 65, 102, 104, 106

Quality objectives, 5

Radiation, 13–14, 17, 27–30, 32, 33
Recycling, 7, 36–40, 47
Revealed preferences, 17, 18, 20
Rhine Commission, 51
River pollution, 3–6
Royal Commission on Environmental Pollution, 29, 30, 75, 88, 91, 97, 99, 100, 101

Society for the Prevention of Asbestosis and Industrial Diseases, 138, 154
Standard setting, 15, 72, 75
Strict liability, 50
Sulphur dioxide, 1, 2, 7, 8, 48, 88
Synergism, 13, 133

Thresholds, 11, 16, 17, 79, 140, 143
Toxic Substances Control Act 1976, 56, 58
Trail Smelter Arbitration ruling 1931, 49
Transfrontier pollution, 2, 48, 49

United Nations Conference on the Human Environment, 49
United Nations Environment Programme, 56, 57

Water authorities, 72, 73, 93, 94